EXPERIMENTAL BIOCHEMISTRY

A Student Companion

EXPERIMENTAL BIOCHEMISTRY
A Student Companion

BEEDU SASHIDHAR RAO Ph.D.

VIJAY DESHPANDE Ph.D.

Department of Biochemistry, University College of Science
Osmania University, Hyderabad

— zinc
— colba
— copper
— acid
sample

Safety Note

It is in the interest of all those working in the laboratory environment to follow and adhere to the lab safety procedures. The methods and protocols described in this book are designed to be carried out by properly trained personnel in a suitably equipped laboratory. Authors have made every attempt to ensure that experimental procedures and information provided in the book are accurate and as per the present standards of laboratory safety. The authors and the publisher cannot be held responsible for incidents and consequential damages of any kind, arising from misuse of the information presented herein.

First published in the UK by

ANSHAN LTD
In 2006

6 Newlands Road,
Tunbridge Wells,
Kent.
TN4 9AT. UK

Tel/Fax: +44 (0) 1892 557767
e-mail: info@anshan.co.uk
Web Site: www.anshan.co.uk

ISBN 1 904798 51 9

© 2005, I.K. International Pvt. Ltd.

British Library Cataloguing in Publication Data
A catalogue record of this book is available from the British Library

Printed in India

This book is dedicated to the beloved memory of

Beedu Raja Rao
(Father of B. Sashidhar Rao)

&

Sandhya Deshpande
(Wife of Vijay Deshpande)

Preface

Developments in the area of modern biology, requires a sound understanding of principles underlying the biochemical & molecular basis of life processes. Investigational approach based on fundamental scientific principles forms the basis of experimental biochemistry. This book is written with a *holistic* approach, encompassing various topics in the practice of experimental biochemistry, so that the information provided is useful to both the student and the teaching community. We have applied a concept-oriented approach in presenting the information in the text, rather than simply providing the facts.

The purpose of writing this book was to bring out an experimental manual that caters to the needs of graduate and post-graduate students, studying in various Indian universities and pursuing Biochemistry as a *de rigueur* subject in their *curriculum*. This book fulfils the requirements of B.Sc., B.Sc. (Hons)., B.Sc. (Pharmacy)., B. Tech.(Biotechnology), MBBS., B.Sc. (Ag) and M.Sc., courses having practical biochemistry module [M.Sc. in Biochemistry, Biotechnology, Microbiology, Environmental science, Food & Nutrition and Life science. (Plant & Animal science)]. Student pursuing a graduate or post-graduate degree course in the area of biochemistry needs a thorough hands-on-training on various aspects of experimental biochemistry as part of their lab *curriculum*. This book is intended to cover a wide variety of *complete experiments* in the area of qualitative & quantitative biochemistry, enzymology, biochemical techniques & preparations, clinical biochemistry, immunochemical methods and food biochemistry. The experimental format includes theoretical considerations, methodological aspects and its application as a practical workout. One of the important approaches in designing this book was to fortify the student's ability to perform an experiment in the laboratory. A *repertoire* of experiments is presented so as to afford a reasonable degree of selection by an instructor. The experimental methods are tailor made to meet the modest facilities available in many of the colleges and universities, within the country.

Foremost emphasis has been given to laboratory safety, as some of the biochemical experiments require the use of hazardous chemicals. Further, a detailed *appendix* has been provided for the convenience of instructors and students.

Comments and constructive criticism from readers would be most welcome.

Department of Biochemistry   *Beedu Sashidhar Rao*
Osmania University   *Vijay Deshpande*
Hyderabad - 500 007, (A.P.).

Acknowledgements

We wish to extend our very special thanks to fellow teachers, who are involved in teaching biochemistry to graduate and post-graduate students in various degree colleges and PG centres in enthusing us to come out with a practical manual that caters, both to the student and the teaching community, as well.

We are grateful to the publishers for their encouragement and support.

Finally, we are thankful to Smt. Bharathi Raja Rao Beedu and Sumitha Rao Beedu, for their goading, patience, assiduous support and extending an unselfish labour, while this book was being written.

Beedu Sashidhar Rao
Vijay Deshpande

Contents

Laboratory Safety and Hygiene

Laboratory safety and health hazards are often overlooked in many laboratories, while performing the experimental work. The purpose of stipulations laid down for laboratory safety and hygiene is to create awareness among the students or the experimenter and to protect them from possible laboratory hazards. It should be noted that, the rules and regulations designed are not meant to scare or to impede the efficiency of the experimenter. Health and safety hazards differ from one laboratory to another, depending upon the type and nature of biochemical experimentation. The term hazard means, the inherent potential of the materials used for experimentation that can cause damage to human health, when handled improperly.

Many biochemical investigations carried out in the laboratory involve use of various components, such as toxic biochemicals, inflammable solvents and compressed gases, potentially infectious biological samples, radioisotopes, fragile glassware, sharp objects and electrical equipment, which pose personal hazards to the experimenter. The laboratory safety guidelines are based on identifying the sources of hazard and by assessing the possible risks associated. A high hazard does not automatically imply a *high risk*, while a low hazard does not mean *no risk*.

Further, special care needs to be taken in disposing of the toxic chemical and biological laboratory wastes generated during the course of experimentation, so that there is no threat to surrounding environment.

This chapter gives an overview of common hazards encountered in the laboratory and some measures to minimise or eliminate them.

Depending upon the nature of experimental work in the laboratory, following potential types of hazards have been identified:

1. Chemical Hazards.
2. Biological Hazards.
3. Physical Hazards.
4. Electrical & Mechanical Hazards.

There are other laboratory hazards which may co-exist along with the above identified hazards. These include, noise, fire, manual handling, penetrating objects etc. It should be noted that, a laboratory experiment can often involve more than one hazard at a time.

CHEMICAL HAZARDS

Chemicals used in the laboratory can present a variety of health and safety hazards. The risks involved here may result from both—storage and use. Chemical hazards can be broadly classified into the following groups (i) *Corrosive chemicals*: These substances can cause injury to the skin or to the body. *Ex*: Phenol, trichloroacetic acid, strong acids and alkalies (ii) *Toxic chemicals*: These agents can cause serious illness or death from exposure to relatively small doses of chemical, either by ingestion, inhalation or absorption through skin. *Ex*: Carbonmonoxide, potassium cyanide, cyanogen bromide. (iii) *Flammable chemicals*: These materials generally have low flash point—below 60°C (the temperature at which the chemical gets ignited) and results in fire, with the release of, heat, smoke, soot and disperse toxic pollutants into the environment. This group of chemicals includes solid, liquid, vapour or gaseous materials. *Ex*: Petroleum products, organic solvents and compressed & liquefied gases (iv) *Highly reactive chemicals*: These chemicals can detonate, explode or liberate poisonous gases upon exposure to light, air, water, oxidizers and other materials. *Ex*: Sodium, ammonium nitrate (v) *Cryogens*: These agents can cause cold burns, frost bite and asphyxia. *Ex*: Liquid nitrogen.

It is mandatory to maintain a file of **M**aterial **S**afety **D**ata **S**heet (**MSDS**) of hazardous substances used for experimental work in the laboratory. Material Safety Data Sheet contains detailed technical information about the nature of hazardous chemical. Normally, as per regulations and stipulations, MSDS should provide the following information:

1. Product identification, common & chemical name.
2. Precautionary labelling.
3. Physical and Chemical characteristics (molecular weight, melting point, boiling point, sp. gravity, flash point, viscosity, solubility, appearance & odour, reactivity, stability, incompatibility, decomposition products, fire & explosion hazard data etc.).

4. Known health hazard data (route of entry, toxicity, effects of overexposure, exposure limits, target organs).
5. Guidelines for safe handling and use.
6. Procedures for clean-up of spills and leakages. Control measures.
7. Transportation data and storage precautions.
8. Emergency first aid procedures.
9. Date of manufacture and Batch #.
10. Name and Address and telephone # of the manufacturer.

BIOLOGICAL HAZARDS

Microorganisms causing biohazard encompass bacteria, *chlamydia*, *rickettsiae*, *mycoplasma*, *protozoans*, fungi and viruses. Improper handling can cause infections and diseases. The routes of potential infection in the laboratory are inoculation, inhalation and ingestion. All specimens of human and animal origin should be regarded and handled as potentially infectious. These include, blood and blood products, urine, faeces, amniotic fluid, tissue samples and cell lines.

PHYSICAL HAZARDS

These include, (i) fire (ii) pressure and (iii) ionizing and non-ionizing radiation.

Fire: Fire is the most potentially devastating emergency in a laboratory that is mediated through a variety of hazardous materials. The fire hazard in the laboratory may originate from ordinary combustible materials such as wood, clothes, plasticware, paper and flammable liquids and gases. The electrical short circuit or naked flame in the vicinity of flammable solvents or sparking due to improperly maintained electrical gadgets may trigger off fire. Alternatively, the fire hazard can also arise from highly reactive chemical agents that are stored improperly.

   Based on the nature of the origin of fire, four categories of fire have been identified (i) *Class A*: This type of fire originates from burning of cellulosic materials, such as paper, wood and synthetic material like rubber, PVC and plastics. Class A type of fire can be extinguished by using water.

(ii) *Class B*: It arises from flammable gases and liquids, which can be extinguished by carbon dioxide or foam type fire extinguisher (iii) *Class C*: This type of fire originates from energised electrical circuits and combustion of materials used therein. Electrical fires are contained by the use of carbon dioxide or dry type chemical fire extinguisher (iv) *Class D*: Combustion of certain metals results in class D fire. These metals include sodium, aluminium and magnesium. For extinguishing class D fire, special, graphite or sodium chloride based dry powder fire extinguisher should be used. Water based fire extinguisher should never be used for class C and D fire.

Pressure: Potential hazards due to high or low pressure systems used in the laboratory are from the use of gas cylinders, pressure vessels (high pressure autoclaves; low pressure thermos or Dewar flask, vacuum desiccators), poor quality flash evaporators, which may result in explosion or implosion.

Ionizing radiation: Ionizing radiation includes, radiations arising from radioisotopes used in tracer technique, such as α-, β-, and γ-radiation. Alpha-emitters are rarely used in biochemical work. Many of the isotopes used in biochemical investigations include β-emitters, example *soft* β-emitters such as ^3H and ^{14}C, *hard* β-emitters such as ^{35}S and ^{32}P and γ- emitters, example ^{125}I, ^{131}I, and ^{60}Co.

Non-ionizing radiation: Non-ionizing radiation is caused by electromagnetic waves, which damage human health at a certain dose of exposure, these include Ultraviolet (200 - 320 nm), LASER's, ELF(extremely low frequency) magnetic fields, static magnetic fields and microwaves.

ELECTRICAL & MECHANICAL HAZARDS

Electrical hazards arise due to the improperly grounded laboratory equipment and electrical appliances. Improper wiring or cable and insulation may also lead to electrical hazard. Other potential sources of electrical hazards include, use of extension cords, heating elements, power sockets, DC (direct current) power supply units used for electrophoresis. Mechanical hazards in the laboratory may originate from the use of instruments and equipment, such as laboratory clinical or high speed refrigerated centrifuges, high speed mechanical tissue homogenizers or macerators and exhaust systems.

STANDARD OPERATING PROCEDURES (SOP) FOR LABORATORY SAFETY

One should remember that, in the laboratory there is no such thing that can be regarded as a harmless substance. Following are the guidelines that are commonly practised as SOP in the laboratory:

▶ Comply with warning signs (hazard symbols) and labels, while working in the laboratory (figure 1.1).

▶ Know the location of Material Safety Data Sheets. Ensure that you have complete information of the chemical that is used for experimental work.

▶ Lab coats or ankle length aprons must be worn while handling toxic, corrosive and flammable materials. Long hair, neckties, or loose clothing should be tied or otherwise secured. Gloves should be worn, while handling corrosive and highly toxic (including allergic, sensitizing) chemicals. Open shoes are not to be worn in the laboratory. Bare legs are not acceptable, while handling hot, cold, toxic, corrosive or sharp materials.

▶ Appropriate eye protection should be worn at all times in laboratories.

▶ Always wash hands with soap after working with chemicals, even though gloves have been used.

▶ Do not mouth pipette or siphon toxic chemical reagents, corrosive liquids, organic solvents, strong acids and alkalies. Use pro-pipette or an auto-dispenser for dispensing.

▶ Do not directly smell, sniff or taste any chemical. Avoid inhalation.

▶ Containers should be closed when not in use.

▶ When working with flammable chemicals, make sure that there are no sources of ignition near by, in order to avoid fire or explosion (like naked flame of Bunsen burner, electrical hot plate etc.).

▶ Handle toxic, corrosive chemicals and flammable solvents in a chemical safety hood or a fume hood.

▶ No smoking in any area of a laboratory.

▶ No eating, drinking of beverage or application of cosmetics in the laboratory, except in designated areas in which no chemicals are used or stored.

▶ Avoid working alone in the laboratory.

▶ If there are any questions about a procedure or the hazards of a chemical, ask the lab supervisor or the instructor before performing the procedure.

▶ Know the location and proper use of emergency equipment (like fire extinguisher, eye wash fountains) and First Aid Kit.

▶ Perform only those experiments or procedures that are authorised by the instructor.

▶ Report all injuries, fires and accidents to the lab supervisor or the instructor, as soon as possible.

HAZARD MANAGEMENT IN THE LABORATORY

1. Management of Chemical hazards:

Chemical Hazards	Consequences	Risk assessment	Hazard management
Flammable chemicals	➤Fire & Explosion	➤Source of ignition ➤Spillage & leakage	Good ventilation. Remove all ignition sources. Proper storage. Emergency procedures. MSDS information.
Corrosive Chemicals	➤Skin & tissue injury	➤Spillage & body contact	Personal protection. Proper storage. Emergency procedures. MSDS information.
Toxic chemicals	➤Systemic poisoning ➤Organ toxicity ➤Carcinogenic ➤Allergic ➤Dermal irritant	➤Route of entry ➤Duration of exposure ➤Environmental & biological monitoring	Good ventilation. Personal protection. Proper storage. Use of safety hood. Emergency procedures. MSDS information.

LABORATORY WASTE DISPOSAL

Clean-up and laboratory waste disposal forms an essential component of good laboratory practice. Different types of wastes are generated during the course of experimentation. These include, broken glassware, empty containers, dispoware (plasticware, needles) chemical, radioactive and biohazardous material (used culture media, contaminated glassware, contaminated agricultural commodities, animal carcass, biological specimens and fluids). Good laboratory disposal practice includes the following guidelines:

▶ Appropriate containers should be provided with labels for the disposal of glass, toxic chemicals, biohazardous materials and flammable wastes.

▶ Laboratory spills should be mopped with an adsorbent material and the area thoroughly cleaned with neutral detergent solution, followed by washing with water.

▶ Small quantities of chemical wastes may be washed down the sink with copious amounts of water. However, this procedure cannot be adopted for toxic chemicals and reagents. Strong acids and alkalies need to be neutralized prior to disposal. Some toxic and organic wastes can be incinerated.

▶ Separate designated area should be identified for carrying out experiments with radioisotopes. Specific containers should be used for the disposal of radioactive waste. The radioactive wastes should be professionally handled under trained manpower (Radiation Safety Officer) with minimum contamination. Highly radioactive material should never be flushed down the drain.

▶ Biohazardous material generated during the course of the experiment should be disinfected before disposal by using strong oxidizing agents like 5% sodium hypochlorite solution or alternatively they may be destroyed by autoclaving or by incineration.

Where there is no proper information about the mode of waste disposal, it is always advisable to consult the instructor or appropriately trained personnel for disposing the laboratory waste.

"There are no circumstances in laboratory experimentation, where the work warrants that significant risks should be tolerated".

Additional Reading

1.   Young, J. (ed), Improving Safety in the Chemical Laboratory – A Practical Guide. Wiley-Interscience, NY, USA, (1987).

2.   Steere, N. (ed), Handbook of Laboratory Safety, 2ⁿᵈ ed. CRC Press, Boca Raton, FL, USA, (1988).

3.   Jadhav, H. V. & Chitnis, M. A. Encyclopaedia of Hazardous Chemicals & Medical Management. Himalaya Publishing House, Mumbai, (1993).

4.   Meyer, E. Chemistry of Hazardous Materials. 3ʳᵈ ed. Brady Prentice-Hall, Inc, NJ, USA, (1998).

Units and Measurements

Experimental biochemistry employs a metric system of units that is based on *Systéme International d'Unités* (SI units). To represent, mass, volume and time biochemists, routinely use terms such as mole, milligram, microgram, parts per million, g%, liter, micro liter, minute, second etc., to express the units in the biochemical analysis. Following are some of the basic units used in practical biochemistry.

MASS

1 g	=	1000 mg (*milligram*)	=	10^{-3} kg
1 mg	=	1000 µg (*microgram*)	=	10^{-3} g
1 µg	=	1000 ng (*nanogram*)	=	10^{-6} g
1 ng	=	1000 pg (*picogram*)	=	10^{-9} g
1 pg	=	1000 fg (*femtogram*)	=	10^{-12} g

Parts per million (ppm)	=	µg per gram or mg per kg	=	10^{-6} g
Parts per billion (ppb)	=	ng per gram or µg per kg	=	10^{-9} g

VOLUME

1 L	= 1000 mL	
1 dL (*deciliter*)	= 100 mL	
1 mL	= 1000 µL (*microliter*)	= 10^{-3} L
1 µL	= 1000 nL (*nanoliter*)	= 10^{-6} L

1mL of water*	= 1.000027 cc (cubic centimeter) = 1 g

* Under most laboratory conditions

CONCENTRATION OF ANALYTE EXPRESSED PER UNIT VOLUME

Molarity (M): Molar concentration of an analyte is the number of moles of solute per one litre (L) of solution. To calculate the molar concentration, the weight of the solute dissolved and its molecular weight are required.

$$\textbf{Moles} = \frac{\text{Weight of solute in grams}}{\text{Molecular weight}}$$

Thus, 1M = 1 mole per liter (**Mole**: The term mole represents the amount of substance (in grams), irrespective of the volume where the substance is present.

Example: (i) IM solution of sodium chloride is prepared by dissolving 58.45 g (MW of Na Cl) of sodium chloride and making up the final volume to one liter with water in a one liter volumetric flask.

$$\text{One mole of sodium chloride} = \frac{58.45}{58.45} = 1$$

(ii) One mole of tyrosine (MW = 181) is 181 g.

(iii) One mole of ovalbumin (MW = 44,000) is 44,000 g or 44 kg.

Usually, the terms such as millimolarity or micromolarity are encountered in biochemical calculations.

1 M	=	1000 mM	=	10^{-3} M	=	1 mole/L
1mM	=	1000 μM	=	10^{-6} M	=	1 mmole/L
1μM	=	1000 nM	=	10^{-9} M	=	1 μmole/L
1nM	=	1000 pM	=	10^{-12} M	=	1 nmole/L

Normality: *Normality* of a solution is the number of gram equivalents of solute per liter of the solution.

$$N = \frac{\text{Mass (gram equivalent) of the dissolved solute}}{\text{Equivalent weight (EW)}}$$

One equivalent of an acid or a base is the mass that contains one mole of replaceable H^+ or OH^- ion (one gram atom or ion).

$$\text{Equivalent weight (EW)} = \frac{\text{Molecular mass}}{N}$$

Where, N is number of replaceable H^+ or OH^- ions per molecule or number of electrons lost or gained per molecule in case of an oxidizing or a reducing agent.

It should be noted that a given solute in solution may have more than one normality value, depending upon the number of electrons lost or gained (oxidation-reduction status) in the reaction. Thus, normality is not a good expression of concentration, as compared to molarity, wherein only one molecular mass exists for a given substance.

The relationship between normality and molarity is given by the following equation:

$$Normality = Molarity \times N$$

Where, N is number of replaceable H^+ or OH^- ions per molecule or number of electrons lost or gained per molecule in case of an oxidizing or a reducing agent.

Example: (i) What is the normality of 18 M sulphuric acid?

Normality of H_2SO_4 = 18 x 2 (number of replaceable H^+ ions in H_2SO_4) = 36

(ii) What is the normality of 5 M NaOH solution?

Number of replaceable OH^- ions in NaOH = 1. Thus, normality of 5 M NaOH = 1 x 5 = 5

CONCENTRATION OF ANALYTE OR SOLUTION EXPRESSED AS PERCENT

(a) *weight / weight*

      g%   = Weight of analyte in g per 100 g of substance.

      mg%   = Weight of analyte in mg per 100 g of substance.

Example: Protein content of groundnut is 25 g % (i.e., 100 g of groundnut by weight contains 25 g of protein).

(b) *weight / volume or volume / volume*

      g% (w/v)   = Weight of solute in g per 100 mL of solvent.

      mg% (w/v)   = Weight of solute in mg per 100 mL of solvent.

Example: (i) 5% (w/v) sucrose solution—Dissolve 5 g sucrose in 100 mL of water (ii) 5% (v/v) methanol solution—Mix 5 mL of methanol with 95 mL of water.

(c) *Percent by volume*

Many a time biochemical analysis requires dilution of % solution (v/v) from higher concentration to a lower concentration. This can be achieved by using the following equation:

Example: Prepare 100 mL of 70% (v/v) ethanol from 95% (v/v) ethanol.

95 (% of ethanol you have) x unknown volume (X) = 70 (% of ethanol you want) x Volume wanted (mL)

$$X = \frac{70 \times 100}{95} = 73.68 \text{ mL of 95\% ethanol.}$$

Thus, to prepare 100 mL of 70% ethanol from 95%(v/v) ethanol - take 73.68 mL of 95%(v/v) ethanol and dilute with 26.32 mL of distilled water.

DENSITY AND SPECIFIC GRAVITY

Density is defined as mass or weight per unit volume, usually given in terms of kg L^{-1} or g mL^{-1}. Specific gravity is a ratio of weight of a substance to the mass of an equal volume of water (density of water at 4°C is 1g/mL). Specific gravity is dimension less, thus numerically specific gravity is equal to density.

Example: Calculate the volume of conc. HCl (35% w/w* ; density 1.18) required to prepare 1L of 0.5 M HCl.

This can be calculated by following equation:

Weight of HCl required = Volume x density (g mL^{-1}) x fraction of HCl weight in 1 mL (weight %)

> = 1000 mL x 1.18 x 0.35 = 413 g
> i.e. One liter of Conc. HCl contains 413 g of HCl by weight.

∴ Number of moles of HCl = Weight / MW = 413/ 36.5 = 11.32
i.e. Conc. HCl is 11.32 M

Thus, to prepare 1L of 0.5 M HCl, the volume of Conc. HCl required is –
0.5 M/ 11.32 M = 0.0442 L, i.e. **44.2 mL**

Take **44.2 mL** of Conc. HCl and make up to 1 L with distilled water in a volumetric flask to give a final concentration of 0.5 M HCl.

IONIC STRENGTH

Ionic strength (*I*) is a measure of the concentration of the total electrical charges existing in solution. It is equal to one half of the sum of the concentration (molarity) of each ion multiplied by square of its valency.

$$I = \tfrac{1}{2} \sum_i M_i Z_i^2$$

Where,

$\sum$ = Sum of . . . .

---

* density of 1mL water = 1 g by weight

$$M_i = \text{Concentration of ion (Molarity)}$$
$$Z_i = \text{Net electrical charge of the ion (+ or -)}$$

The relationship between the molarity and ionic strength depends on the concentration or number of ionizable species (expressed as molarity) in the solution and their net electrical charge.

Example: (i) What is the ionic strength of 150 mM solution of sodium chloride?

Ionic strength is calculated by the equation - $I = \frac{1}{2} \sum M_i Z_i^2$

$$150 \text{ mM NaCl} = 0.15 \text{ M}$$
$$\text{Ionization of NaCl gives - Na}^+ \text{ and Cl}^-$$

Ionic strength of NaCl = $\frac{1}{2} \sum [(0.15) \times (1)^2 \text{(for Na}^+ \text{ion)}] + [(0.15) \times (1)^2$ (for Cl$^-$ ion)]

$$= \mathbf{0.15}$$

(ii)   Calculate the ionic strength of 0.6 M Na_2HPO_4 solution.

Substituting the values in the equation .... $I = \frac{1}{2} \sum M_i Z_i^2$,

The 0.6 M Na_2HPO_4 solution yields - $2Na^+$ (0.4 M) and
HPO_4^{2-} (0.2 M) ions

Ionic strength of 0.6 M Na_2HPO_4 solution = $\frac{1}{2} \sum (0.4) \times (1)^2 + (0.20) \times (2)^2 = \mathbf{0.6}$

*[See, **appendix** for (i) Molarity, normality, % conc., & specific gravity of routinely used acids & alkalies in the laboratory (ii) Density & boiling point of few organic solvents].*

Additional Reading

1.   Montgomery, R., & Swenson, C. A. Quantitative Problems in the Biochemical Sciences. W. H. Freeman & Co. San Francisco, USA, (1969).

2.   Dawes, E. A. Quantitative Problems in Biochemistry. 6th ed. Longman, London, UK, (1980).

3

Basic Statistical Concepts for Biochemical Analysis

Results obtained in any biochemical investigations are in the form of numerical data. Inference from such numerical data can be drawn only after scientifically organising, summarizing, analysing and presenting the data. Application of statistical methods for collecting, analysing, interpreting and presenting biological measurements (quantitative information) is termed as *Biometry* or more often referred to as *Biostatistics*.

The working knowledge of biostatistics is essential for a biochemist, for the assessment of experimental data with respect to error analysis. Error is either (i) the difference between an experimental result or (ii) the degree of uncertainty of an experimental result. Two commonly used statistical terms involved in error analysis are,

        (i)   Reproducibility or *Precision.*
       (ii)   Repeatability or *Accuracy.*

PRECISION

This term is a measure of reproducibility or reliability of an experimental procedure or technique. In terms of biochemical data, it indicates the closeness of the number of replicate measurements of a defined biological sample. Statistically, it is a measure of *random error*. Precision of an experimental investigation is expressed in terms of standard deviation (σ) or % coefficient of variance (% CV).

ACCURACY

Accuracy is a measure of the nearness of an experimental result or the mean value of replicate analyses to the true value for a given biological sample. Normally, smaller the systematic deviation of an experimental error, more accurate is the experimental procedure. Statistically, it is a measure of systematic error or *bias* and is usually being reported as % bias.

$$\% \text{ bias} = \frac{\text{measured value} - \text{true value}}{\text{true value}} \times 100$$

While working with biological samples, the true value is not known, making it difficult to determine the extent of experimental bias.

For a biochemist, the determinate errors (which includes both, random errors and most of the systemic errors) are those errors that can be controlled by the experimenter. The usual sources of the errors are associated with (i) faulty equipment (ii) improperly designed experiment (iii) poorly controlled experimental conditions and (iv) biochemical interference.

The error component of a biological experiment can be corrected or reduced considerably by using calibrated equipment, performing an experiment under properly controlled experimental conditions and using a carefully planned experimental design. The indeterminate errors or accidental errors of an experiment cannot be controlled or evaluated by an experimenter.

The concept of precision and accuracy is illustrated diagrammatically (Fig. 3.1). In this figure the concentric circles represent the targets, while

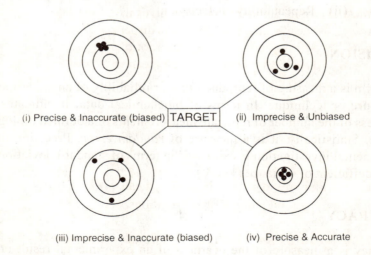

(i) Precise & Inaccurate (biased) |TARGET| (ii) Imprecise & Unbiased

(iii) Imprecise & Inaccurate (biased)   (iv) Precise & Accurate

Figure 3.1 Analogy for precision and accuracy

the black dots on the target are the shots fired from a gun. In this analogy, each gunshot represents a replicate observation. The observations can be summarised as follows (i) all the gunshots are grouped together outside the bullseye of the target—indicating that the gunshots were precise, and inaccurate (biased), as they had missed the bullseye; (ii) the gunshots are spread out in the inner circle of the target -indicating that shots to be unbiased but imprecise; (iii) the gunshots are distributed between the bullseye and the concentric circles of the target—indicating that the gunshots were imprecise and inaccurate (biased) and (iv) the gunshots are grouped on the bullseye— indicating that the shots were precise and accurate.

COMPUTATION OF MEAN, STANDARD DEVIATION, COEFFICIENT OF VARIATION AND STANDARD ERROR

Mean: It is a measure of central location, providing a value around which a set of data is located. Mean of n items is obtained by dividing the sum of n items by the number of them.

$$\text{Mean } \bar{x} = \frac{x_1 + x_2 + x_3 + \ldots\ldots + x_n}{n}$$

Standard deviation: Standard deviation gives the dispersion of numerical data around the mean value. The greater the dispersion, larger is the standard deviation. By definition, it is the square root of the mean of the squared dispersions of the measurement from their mean value. It is calculated using the following equation:

$$\text{S.D.} = \sqrt{\frac{(x - \bar{x})^2}{n}} = \sqrt{\frac{\sum d^2}{n}}$$

Where, x = value of the observation; $\bar{x}$ = mean value of the observation; d = deviation; n = total number of observations.

Coefficient of variation: This is a measure of relative variability. It is a product of coefficient of standard deviation x 100.

$$\text{Coefficient of variation (CV)} = \frac{\text{Mean}}{\text{S.D.}} \times 100$$

Standard error: Often, while presenting the probable error of the biological data, the term standard error or standard deviation of the mean is used. This is calculated by the following equation,

$$\text{S.E.} = \frac{\text{S.D.}}{\sqrt{n}}$$

The precision of a biological measurement can be improved by using the relationship between the number of experimental observations and standard deviation of the mean. In order to halve the value of standard error, the number of experimental observations has to be increased by four times.

Example 1: The protein content of five cultivars of groundnut was estimated to be 22, 24, 25, 28 and 26 gram percent. Calculate the mean protein content and its deviation or variability, among the groundnut cultivars.

$$\text{Mean protein content} = \Sigma \frac{22+24+25+28+26}{5} = \frac{125}{5} = 25 \text{ g\%}$$

Calculation of standard deviation:

Cultivar #	Protein content (g%)	Deviation = (x - mean value) = (x - 25)	$d^2 = (x - mean)^2$
1	22	- 3	9
2	24	- 1	1
3	25	0	0
4	28	+3	9
5	26	+1	1
		$\Sigma d = 0$	$\Sigma d^2 = 20$

$$S.D. = \frac{\sqrt{\Sigma(x-\bar{x})^2}}{\sqrt{n}} = \frac{\sqrt{20}}{\sqrt{5}} = \mathbf{2.0}$$

Comments: The mean protein content of five cultivars is 25 g%. The variability in the protein content among the groundnut cultivars is ± 2.0g%. Thus, the experimental data is presented as **Mean ± S.D**, i.e., 25.0 ± 2.0 g%.

Example 2: A group of ten biology graduate students analysed blood glucose level by glucose oxidase method, after aliquoting a sub-sample from a given human blood sample. The glucose concentrations reported by them were as follows, 89, 95, 98, 98, 104, 95, 90, 96,100 and 95 (mg dL⁻¹). Calculate the mean, S.D., CV and S.E. from the given data.

Tabulating the data,

Student code No.	Glucose conc. (mg dL⁻¹)	Deviation = $(x$ - mean value $)$ $(x - 96)$	$d^2 = (x - \text{mean})^2$
1	89	- 7	49
2	95	- 1	1
3	98	+2	4
4	98	+2	4
5	104	+8	64
6	95	-1	1
7	90	- 6	36
8	96	0	0
9	100	+4	16
10	95	-1	1
$n = 10$	960	$\Sigma d = 0$	$\Sigma d^2 = 176$

(i)   Mean ($\bar{x}$) =   960/10   =   **96**

(ii)   S.D.   = $\dfrac{\sqrt{\Sigma d^2}}{\sqrt{n}}$ = $\dfrac{\sqrt{176}}{\sqrt{10}}$ =   **4.2**

(iii) CV   = $\dfrac{S.D.}{\bar{x}} \times 100$ = $\dfrac{4.2}{96} \times 100$ =   **4.4%**

(iv)   S.E   = $S.D./\sqrt{n}$ = $4.2/\sqrt{10}$ =   **1.33**

Comments: The mean and standard deviation of the blood glucose concentration in the sample is 96 ± 4.2 mg dL^{-1}. The coefficient of variation and standard error is 4.4% and 1.33, respectively. The variation in the reported values of glucose may be attributed to indeterminate and determinate errors (possibly, instrumental, pipetting or human error).

Additional Reading

1.   Eckschlager, K. Errors, Measurement and Results in Chemical Analysis. Van Nostrand Reinhold Company, London, (1972).

2.   Hannagan, T. J. Work Out Statistics. 2nd ed. Macmillan Education Ltd. London, UK, (1987).

3.   Day, R.A. Jr., & Underwood, A.L. Quantitative Analysis. 6th ed. Prentice-Hall of India (P) Ltd. New Delhi, (1993).

4.   Arora, P. N. & Malhan, P. K. Biostatistics. Himalaya Publishing House, Mumbai, (1998).

5.   Holme, D. J. & Peck, H. Analytical Biochemistry. 3rd ed. Addison Wesley Longman Ltd. Essex, UK, (1998).

Basic Concepts and Use of Instruments in Biochemical Analysis

1. PHOTOMETRY AND SPECTROPHOTOMETRY

Photometry is one of the most widely used spectroscopic methods for the quantitative analysis of biomolecules. Photometry and spectrophotometry involves the interaction of light energy (electromagnetic radiation, having both wave and particle or photon form) with matter. The electromagnetic spectrum includes ultraviolet (UV), visible and infrared regions. Various regions of the electromagnetic spectrum are depicted in figure 4.1. Energy content in the various regions of the spectra decreases with the increase in the wavelength ($1 - 1 \times 10^6$ nm). The spectra corresponding to UV-visible region is of particular interest to biochemists. When wavelengths of radiation impinge on solution containing molecules, the radiant energy in the form of photons is transferred to the absorbing molecules. This absorption depends upon the electronic structure of the absorbing molecules. Measurement of the amount of incident radiation absorbed will provide quantitative information about the number of absorbing molecules.

Physical laws of light absorption

The absorption of light by a given solute or a biomolecule in solution is governed by three important factors (i) the wavelength of the incident radiation (ii) the concentration of the solute and (iii) the length of the incident light path in solution. Extensive experimental studies in relation to the interaction of above factors resulted in the formulation of laws governing quantitative absorptiometry.

Lambert's law: This law states that the intensity of a parallel beam of transmitted monochromatic radiation diminishes exponentially with the thickness or the length of the absorbing medium.

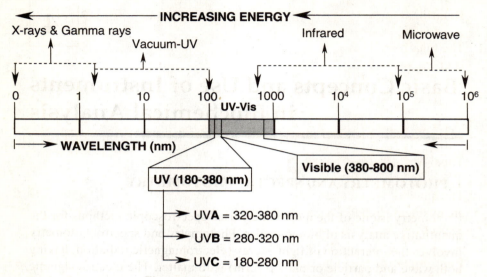

Figure 4.1 The electromagnetic spectrum

Beer's law: This law states that the absorbance at any given wavelength varies directly to the concentration of the absorbing substance in solution. Further, the fraction of the transmitted radiation through the solution decreases exponentially with the increasing concentration of the absorbing substance.

Beer-Lambert's law is the fundamental law on which quantitative measurements in colorimetry and spectrophotometry are based. This law states that the amount of incident radiation absorbed or transmitted by a solution or medium is proportional to the total concentration of absorbing molecules present in the light path. This relationship is given by the following equation,

$$A = \varepsilon\, l\, c$$

Where,

A	=	Absorbance or $(-\log I / I_o)$.
I_o	=	Intensity of the light impinging on the sample.
I	=	Intensity of the light transmitted through the sample.
ε	=	Molar absorption coefficient.
l	=	Length of light path through the sample or thickness of the sample cell.
c	=	Concentration of the absorbing substance in the sample.

Relationship between Absorbance (A) and Transmittance (T)

The logarithmic function of I and I_o expressed in terms of % transmittance is given by the following equation,

$$\%T = I / I_o \times 100 \qquad\qquad ...(Equation\ 1)$$

$$A = - log\ (I / I_o) = - log\ (\%T/100) = log\ (100\ /\ \%T) \quad ...(Equation\ 2)$$

Thus, Absorbance $(A) = (2 - log\%T)$

The quantitative measurements of absorption of radiation by biomolecules are evaluated experimentally by using instruments such as a photometer or a spectrophotometer.

Photometer & Spectrophotometer

A photometer is an instrument which measures the intensity of a light beam in the visible region of the spectrum (usually in the region of 400 - 750 nm). This instrument uses coloured filters (usually made of glass or gelatin) and a simple photoelectric detector to measure the transmitted light. The photoelectric device converts transmitted light energy into electric current, which is recorded analogally or digitally. Often, this instrument is also referred to as photoelectric colorimeter that is used routinely for colorimetric estimations in biochemistry and clinical laboratories.

A spectrophotometer is a sophisticated instrument used to measure the transmitted light in the spectral regions of UV and visible range. The advantages of using a spectrophotometer over the photometer are (i) presence of monochromatic light source (both UV and visible light sources) (ii) increased sensitivity (iii) wide specʋal range (iv) narrow spectral regions (v) recording the absorbance of dilute solutions and (vi) facility to record— absorption spectra of substances as a function of the wavelength.

Important components of a spectrophotometer include: (i) a radiant energy source (ii) monochromator (iii) sample chamber (iv) photodetector system and (v) recorder or a display device. Figure 4.2 illustrates the various components of a conventional spectrophotometer.

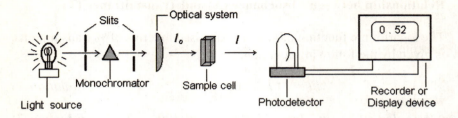

I_o = **Intensity of incident light** I = **Intensity of transmitted light**
Light source: UV region = Deuterium discharge lamp (190 - 340 nm range)
                      Visible region = Tungsten-halogen lamp (340 - 800 nm range)
Monochromator = Prism or Diffraction grating or Holographic grating.
Sample cell = Quartz or Fused silica cuvette for UV-Visible region; Glass cuvette
for Visible region.
Photodetector: Photomultiplier tube (PMT) or Photodiodes

Figure 4.2 Basic components of a spectrophotometer

Radiant energy source

Spectrophotometer has two sources of radiant energy, one for the UV region and the other for the visible region of the electromagnetic spectrum. High pressure deuterium discharge lamp is used as a source of UV radiation. This lamp produces radiation in the region of 190-340 nm. Usually tungsten-halogen lamp is used as a source of light for the visible region (340-800 nm).

Monochromator

Monochromator is a device that divides the incident light beam into its component wavelengths i.e., monochromatic light of a specific wavelength. Monochromatic light is achieved by using a prism, or a diffraction grating or a holographic grating. A grating is a parallel array of equidistant grooves, closely spaced (~2000 grooves per mm), which disperse the polychromatic light into its component wavelengths. Usually glass or metal mirrors are used as the media for making the grating. The spectral purity of the monochromatic beam is increased by passing the monochromatic light through a series of slits, lenses, filters and mirrors (collectively referred to as optical system of a spectrophotometer).

Sample cell

Spectrally pure monochromatic light is directed into a sample chamber, which accommodates a sample cell or cuvette. Cuvettes, made out of glass, quartz or fused silica, are used as sample cells. Spectrophotometric measurements of biological samples to be evaluated are taken in these cuvettes. Quartz cuvette is used for UV and visible spectrometric measurements, while glass cuvette is used for visible regions (400-800 nm) only. As glass cuvette absorbs UV light, they cannot be used for evaluating the biomolecules that absorb radiant energy in the UV region.

Photodetector system

A photoelectric detector produces current which is proportional to the light impinging on it. Photomultiplier tube (PMT) is a very sensitive photoelectric device which detects very small amounts of light energy. The PMT consists of a series of electrodes. The incident light falling on the first electrode causes the release of electrons that are subsequently amplified and accelerated by dynodes into electrical signals. A solid state device namely, the photodiode detectors are presently being used in modern spectrophotometers. These are silicon-based devices that are sensitive to photons in the wavelength range of 170 to 1100 nm. The electrical signals generated by the PMT device or photodiodes are recorded analogally or digitally.

Recorder or display device

The electrical signals generated by the photoelectric detector are recorded as absorbance or % transmittance by analog meter or a digital display unit. Alternatively, the signals can also be processed by a computer.

Applications

Important biochemical applications of spectrophotometer include: (i) measurement of absorbance of a biomolecule in solution at a fixed wavelength, which is useful in obtaining the quantitative data of a given biomolecule (ii) determination of the absorption spectra of a given compound over a range of wavelengths i.e., measurement of absorbance as a function of wavelength (λ). Absorption spectra provide qualitative data with respect to the identity and structure of the biomolecules (iii) determination of the molar absorption coefficient of biomolecules. Figures 4.3 & 4.4 depict the UV- absorption spectra of (i) a protein (bovine serum albumin) and (ii) Calf thymus DNA sample, respectively. Absorbance of these biomolecules is plotted as a function of wavelength.

Examples

(i) *Computation of the concentration of an analyte in a sample solution using Beer-Lambert's law:* A sample solution of L-tyrosine, dissolved in 10 mM sodium carbonate solution and taken in a 1 cm cuvette, showed an absorbance (A) of 0.9, at a wavelength of 295 nm. The molar absorption coefficient of the amino acid is 1500 M^{-1} cm^{-1}. What is the molar concentration of the amino acid?

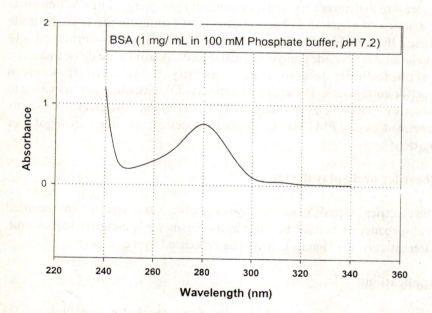

Figure 4.3 UV - Absorption spectra of a protein (BSA)

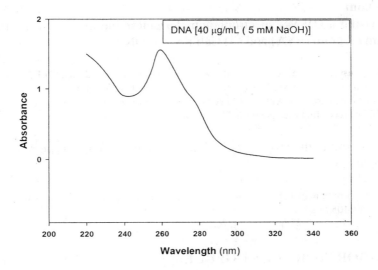

Figure 4.4 UV - Absorption spectra of Calf thymus DNA

$$A = \varepsilon lc$$

$A = 0.9;$ $\qquad l = 1\,cm;$ $\qquad \varepsilon = 1500 \text{ M}^{-1} \text{ cm}^{-1}$

$c = A/\varepsilon l,\ 0.9 \div 1500 \text{ M}^{-1} \text{ cm}^{-1} \times 1 \text{ cm} = 0.0006 \text{ M or } \mathbf{6 \times 10^{-4} \text{ M}}$

The concentration of L-tyrosine in the sample solution is **6 x 10⁻⁴ M**.

(ii) *Determination of molar absorption coefficient (ε):* The absorbance (*A*) of a 7 x 10⁻⁵ M solution of guanine (dissolved in 0.1N HCl) at a λ_{max} 275 nm is 0.56. The pathlength of the cuvette is 1 cm. Calculate the molar absorption coefficient of guanine (*ε*).

$$A = \varepsilon lc$$

$A = 0.56;$ $\quad c = 7 \times 10^{-5} \text{ M}$ $\qquad l = 1\,cm;$ $\qquad \varepsilon = ?$

$\varepsilon = A/cl,$ $\qquad 0.56 \div 7 \times 10^{-5} \text{ M (mole/L)} \times 1 \text{ cm} = 8 \times 10^3 \text{ M}^{-1} \text{ cm}^{-1}$

The molar absorption coefficient (*ε*) of guanine = **8 x 10³ M⁻¹ cm⁻¹**

Workout:

(**A**) Using a UV-Vis spectrophotometer determine the λ_{max} of (i) Bovine serum albumin (BSA) and (ii) Calf thymus DNA.

Note: (i) Dissolve protein (1mg/mL) in 100 mM sodium phosphate buffer, *p*H 7.2 (ii) Dissolve DNA (1mg/mL) in 5 mM NaOH. Dilute 1:20 with 5mM NaOH and use.
Calibrate the instrument with the respective blanks, before recording the spectra. Use quartz cuvette for recording the spectra in UV region.

(**B**) Determine the molar absorption coefficient of (i) L-tryptophan (ii) *p*-nitrophenol.

Note: Tryptophan in 0.1M NaOH has λ_{max} of 280 nm; *p*-nitrophenol in 0.1N NaOH has λ_{max} of 405 nm.

2. LABORATORY CENTRIFUGE

A centrifuge is an electro-mechanical laboratory equipment that is routinely used in a biochemistry laboratory for sedimentation and separation of suspended solids from a liquid. Another valuable application is for the separation of liquids with different specific gravities and two immiscible liquid phases. The important components of a laboratory centrifuge include, (i) high speed electrical motor (ii) rotor (iii) speed regulator and (iv) tachometer or digital speed indicator.

The electrical motor along with the drive shaft is the most important component of the centrifuge. The rotor of the centrifuge holds the sample tubes in position and is normally made of aluminium alloy or brass. The surface of the rotors is either anodized or epoxy painted. They are mounted on the drive shaft of the motor. Generally, two types of centrifuge rotors are commonly used in biochemistry laboratories that are referred to as (i) horizontal head or swinging bucket type and angle head or fixed-angle type. Angle head rotors can attain much higher speeds than horizontal head rotors due to its very low friction with air while in motion. At high speed, air resistance results in heat generation due to friction. Many high speed centrifuges are equipped with refrigeration system to avoid warming of the sensitive biological samples. In addition, a centrifuge is provided with a tachometer or digital speed indicator to indicate the speed attained by the motor, which is displayed as revolutions per minute (*rpm*). The speed of the centrifuge is controlled by potentiometer which raises or lowers the voltage applied to the motor. It should be noted that, the relative voltage increment can never be taken as an accurate indicator of speed. The term *rpm* when applied to a centrifuge does not give any indication of the applied centrifugal force on the sample.

Fundamental concepts of centrifugation

Centrifuges are designed to accelerate sedimentation of particles, uniformly suspended in solution, by taking advantage of applied centrifugal force. The impact of this force is larger than the earth's gravitational field, thus increasing the rate at which the particle sediments. The sedimentation rate of the particles depends on the density, molecular mass, shape and the viscosity of the liquid medium. When a centrifugal force is applied to such suspended particles, the rate at which they sediment is directly proportional to the intensity of the applied centrifugal field. The centrifugal force F generated during the course of centrifugation is defined by,

$$F = m\omega^2 r$$

Where,

F = intensity of the applied centrifugal force.
m = effective mass of the sedimenting particle.
ω = the angular velocity of the rotation in radians s^{-1}.
r = the distance in cm of the sedimenting particle from the central axis of the rotor.

The force acting on a sedimenting particle increases with the speed of the rotation and distance of the particle from the axis of rotation. A more practical measurement of F in terms of earth's gravitational field is represented as *relative centrifugal force (RCF)*, defined by the equation,

$$RCF = (1.118 \times 10^{-5})\,(rpm)^2 \times r$$

Relative centrifugal force is expressed as some number x g. To illustrate this, consider a sample in a rotor with an apparent radius of 4 cm being rotated at 5000 rpm, can be represented as 11800 x g, applying the above equation.

A wide range of centrifuges ..re available in market for various applications. They may be classified broadly into three categories (i) low

Table 1: Types of centrifuges and its applications.

Type of centrifuge	Maximum speed attainable (rpm)	Maximum RCF attainable	Refrigeration	Applications
Low-speed (Bench top model).	8000	6000 x g	Some	Clinical biochemistry, Cell and nuclei separations.
High-speed (Bench top or Floor model).	25000	18000 x g	Yes	Separation of cells and cellular organelles.
Ultra-high speed (Floor model).	80000	6×10^5 x g	Yes	Separation of bio-macromolecules.

speed (ii) high speed and (iii) ultra-high speed. Table 1 gives different types of centrifuges and their applications.

Care and handling of centrifuge

Laboratory centrifuge is potentially a dangerous instrument, if used improperly. While, using the equipment, it is essential to follow the directions precisely, such as (i) the load in the centrifugal cups should be perfectly balanced (ii) ensure that the glass centrifuge tubes, do not project so far as to strike the centre of the head, when it is swung in its place and is in the horizontal position (iii) while turning on the power, ensure that the speed controller (potentiometer) is at zero position (iv) the power supply to the motor should be increased slowly, until the desired *rpm* is attained for designated load (v) if an unusual noise or vibration is observed during centrifugation, turn off the instrument immediately to avoid any further damage. These simple tips, ensure the safety of the user, life of the centrifuge and success of the experiment.

3. ELECTROPHORESIS

Electrophoretic technique is one of the principal biochemical tools employed in the analysis of biomolecules possessing ionizable groups. These molecules include amino acids and their derivatives, peptides, proteins and nucleic acids.

Electrophoresis involves movement of charged particles in an electrical field. To illustrate this, an amino acid exists as a positively charged species (cation) in acidic media, while in alkaline condition as negatively charged ion (anion) (fig. 4.5). Similarly, proteins contain numerous amino (basic) and carboxyl (acidic) groups, which undergo ionization depending upon the *p*H of the solution. In acidic condition, they assume a positive charge and exhibit cathodic mobility. On the contrary, at alkaline *p*H, proteins acquire a negative charge and show an anodic migration.

The net charge on the protein depends upon the ionization of the total number of amino and carboxyl groups. Greater the net charge (positive or

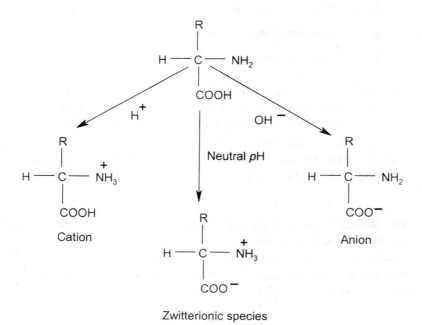

Figure 4.5 pH dependent ionization of an amino acid.

negative charge) on the protein, greater will be its mobility under the applied electrical field. Further, molecules having similar net charge but differing in their molecular mass, exhibit inherent differences in their charge to mass ratio, under applied electrical field. This property is exploited for effective separation of charged species by electrophoresis.

Physical laws governing electrophoresis

The electrophoretic separation of biomolecules is governed by the following equation,

$$V = \frac{Eq}{f} \qquad \ldots \text{equation 1}$$

Where,

V = Velocity at which the charged species moves

E = Electrical field in Volts/cm or the potential gradient

q = The net charge on the molecule

f = The frictional coefficient

In simple terms, the mobility of the charged species is directly proportional to the applied voltage or the net charge present on the molecule. Alternatively, mobility is inversely proportional to the friction or the drag of the molecule during the electrophoretic migration. Experimentally, it is arduous to determine the f value. Thus, the equation 1 is inadequate to describe the process of electrophoresis.

Factors influencing electrophoretic mobility: The flow of current between the electrodes is mediated mainly by the buffer ions, while the sample ions contribute to a small proportion to this electrical conductance. The relationship between the current in mA (I), voltage (V) and resistance (R) in Ohms is given by the following equation:

Ohm's law, $V = IR$

$$\text{or resistance } (R) \text{ is defined as} = \frac{V}{I}$$

Thus, an increase in either current or voltage will lead to increase in migration of charged species. Increase in the applied voltage or current results in heat generation and consequently rise in temperature because of molecular friction due to the mass and shape of the molecule, in the separating media. The increase in temperature leads to decrease in resistance and if not controlled, would affect the electrophoretic separation. Electrophoretic separation is also influenced by the duration (*time*) of the run. Greater the time of migration, greater will be the electrophoretic separation of the sample components.

The ionic strength and *p*H of the buffer medium used during the course of electrophoresis, maintains a constant ionization environment for the molecules under separation. Any change in this environment alters the net charge and the rate of migration during electrophoresis. Heating of the electrophoretic medium ensues the following effects (i) formation of convectional currents leading to mixing up of separated components (ii) increases the rate of sample diffusion resulting in broadening of the separated species and (iii) decrease in viscosity of the buffer.

Electrophoretic equipment

It consists of (i) power supply unit / power pack, that supplies *Direct Current*. Normally, power packs have provision for both constant current and constant voltage supply (ii) an electrophoretic apparatus made of perspex or acrylic material, consisting of cathodic and anodic compartments along with platinum electrodes. The apparatus has terminals for connecting the unit to power pack through red and black colour coded power cords.

Various types of electrophoretic systems have been developed for numerous applications in biochemical analysis. These include, paper, cellulose acetate, polyacrylamide (tube and slab gel) and agarose gels.

*(See, **appendix** for (i) Physical constants, coloured filters & complementary hues, molar absorption coefficient of selected biochemicals and (ii) Determination of nucleic acid and protein concentration by UV-absorption method).*

Additional Reading

1.   Robert, H. Jerald, R. S. Mary, A. S. & McWhorter, C. A., (eds). Laboratory Instrumentation. Harper & Row Publishers, Maryland, USA, (1974).

2.   Lee, L. W. & Schmidt, L. M. Elementary Principles of Laboratory Instruments. 5th ed. The C. V. Mosby Company, St. Louis, USA, (1983).

3.   Christian, G. D. & 'O' Reilly, J. E. (eds), Instrumental Analysis. 2nd ed. Allyn & Bacon. Inc. Boston, USA, (1986).

4.   Cooper, T. C. Tools of Biochemistry, John Wiley, New York, USA, (1997).

5.   Hawcroft, D. M. Electrophoresis the Basics. IRL Press, Oxford, UK, (1997).

6.   Wilson. K. & Walker, J. Principles and Techniques of Practical Biochemistry. 4th ed. Cambridge University Press. UK, (1995).

pH and Buffers

All biochemical reactions in cells and tissues eventuate in aqueous *milieu*. These reactions depend upon the rigid regulation of hydrogen ion activity, in *in vivo* condition. In order to understand these biochemical processes *in vitro*, it is necessary to *mimic* these conditions in artificial media, wherein hydrogen ion concentration plays a portentous role. The term pH denotes the hydrogen ion concentration or activity of a given solution. The symbol 'p' denotes the *negative logarithm of*, while '**H**' represents the *hydrogen ion* concentration. It is defined by the following expression:

$$pH = \log 1/[H^+] = - \log [H^+]$$

One of the important properties of pure water is its *amphoteric* nature. It serves as a proton donor (acid) and acceptor (base). This is represented by the following equation of ionic equilibrium of water,

$$H_2O \leftrightarrow H^+ + OH^-$$

$$H_2O + H^+ \leftrightarrow H_3O^+ \ (hydronium \ \text{ion}).$$

In pure water, the degree of ionization at equilibrium is small. For every 10^7 molecules of water, one molecule is in ionized form at 25°C. The ion product of $[H^+] [OH^-]$ in aqueous solution (at 25°C) always equals $1 \times 10^{-14} \ M^2$. This is based on the following expression:

$$K_w = [H^+][OH^-] = (55.5 \ M) \ (1.8 \times 10^{-16} \ M) = \mathbf{1.0 \times 10^{-14} \ M^2}$$

Where, (i) K_w is ion product of water at 25°C.

(ii) 55.5 M is the concentration of water at 25°C (calculated by dividing weight of one liter water in grams by gram molecular weight, i.e., 1000 g ÷ 18.02 = 55.5 M).

(iii) 1.8×10^{-16} M is the value of K_{eq} {equilibrium constant for reversible ionization of water - K_{eq} = [H$^+$] [OH$^-$]/[H$_2$O]} based on the electrical conductivity measurement of pure water.

Thus at neutral *p*H the concentration of H$^+$ and OH$^-$ ions are exactly equal. This is arrived by the following equation,

$$K_w = [H^+][OH^-] = [H^+]^2 \ldots . \ Equation \ 1.$$

Solving the equation 1 for [H$^+$] gives,

$$[H^+] = \sqrt{K_w} = \sqrt{1.0 \times 10^{-14} \ M^2}$$

$$[H^+] = [OH^-] = 10^{-7} \ M$$

Equation 1 is useful in calculating the [H$^+$] in solutions.

Example 1. Calculate the hydrogen ion concentration in 0.1M potassium hydroxide solution.

Using the equation 1: K_w = [H$^+$][OH$^-$] = [H$^+$]^2 = K_w / [OH$^-$] = 1.0×10^{-14} M^2/0.1M = **10^{-13} M**.

*p*H SCALE

Ion product of water forms the basis of *p*H scale (Fig. 5.1), that can be derived from the following relationship of *p*H and *p*OH,

$$K_w = [H^+][OH^-] \ldots .. \ \text{taking log of this equation}$$

$$- \log [H^+] + \log [OH^-] = \log K_w$$

$$- \log [H^+] = pH; \qquad - \log [OH^-] = pOH; \qquad - \log K_w = pK_w$$

$$\therefore pH + pOH = pK_w$$

$$K_w = 1 \times 10^{-14} \ pK_w = - \log 1 \times 10^{-14} = +14$$

$$\mathbf{pH + \textit{p}OH = 14}$$

$[H]^+$	$10^0(1)$	10^{-1}	10^{-2}	10^{-3}	10^{-4}	10^{-5}	10^{-6}	10^{-7}	10^{-8}	10^{-9}	10^{-10}	10^{-11}	10^{-12}	10^{-13}	10^{-14}
pH	0	1	2	3	4	5	6	7 (Neutral)	8	9	10	11	12	13	14
pOH	14	13	12	11	10	9	8	7 (Neutral)	6	5	4	3	2	1	0
$[OH]^-$	10^{-14}	10^{-13}	10^{-12}	10^{-11}	10^{-10}	10^{-9}	10^{-8}	10^{-7}	10^{-6}	10^{-5}	10^{-4}	10^{-3}	10^{-2}	10^{-1}	$10^0(1)$

Fig. 5.1 The *pH* scale.

Note: $pH + pOH = 14$

Measurement of *p*H is one of the routinely employed procedures in biochemical analysis. Some of the important interactions of biological significance like catalytic activity of enzymes, structure-activity relationship of biomacromolecules are *p*H dependent.

*p*H METER

One of the routinely used instruments in a biochemistry laboratory is the *p*H meter. The *p*H meter is a device designed to measure the effective concentration of hydrogen ions in solutions. A typical *p*H meter measures the potential difference (electromotive force - EMF) that develops between a combination ion selective glass electrode and an unknown solution containing hydrogen ions. The magnitude of this EMF is also dependent on the temperature of the solution. The output potential is in millivolts (mV) and is recorded galvanometrically or digitally with a scale graduated in *p*H units. This relationship is defined by the following equation:

$$V = \frac{E_o + 2.303\, RT}{F} = p\text{H}$$

Where,

V = voltage of the completed circuit (observed *emf*).

E_o = potential of reference electrode.

R = the gas constant (8.341 J/mole per °K).

T = the absolute temperature in °K (25°C = 298°K).

F = the Faraday's constant (96,4846 Coulombs/ equivalent weight or 9.64846×10^4 Coulombs mol^{-1}).

The combination electrode consists of a reference electrode and a *p*H sensitive glass electrode, connected by KCl salt bridge, contained in a single glass tube. This electrode develops 59.15 mV per *p*H unit at 25°C and this out put increases linearly to 74.05 mV per *p*H unit at 100°C. In addition, the *p*H meter contains (i) an electronic amplifier that is necessary for the measurement of potential generated in the glass electrode, as extremely small current will flow in the circuit due to high internal resistance of the glass electrode. This amplifier unit has high input impedance to ensure high accuracy and performance (ii) an offset potentiometric control for the calibration of the instrument with external reference buffer solution and (iii) a rotary mode switch that activates the selection of pH or mV measurement mode. Modern pH meters have built in automatic temperature compensator (ATC) for temperature setting.

In practice, the pH meter is calibrated before use by standardizing it against the reference standard solutions of known pH. Traditionally, 0.05M potassium hydrogen phthalate is used as reference standard for acidic range (pH 4.01 at 30°C) and 0.01M borax for alkaline range (pH 9.14 at 30°C). Potassium-sodium phosphate buffer (0.05M) is used as reference standard for neutral pH (pH 6.85 at 30°C).

BUFFERS

A buffer is defined as a chemical system that resists change in pH upon the addition of small amounts of hydrogen or hydroxyl ions. It plays a crucial role in the maintenance of near constant pH, which is essential for numerous biochemical reactions occurring in biological systems.

Commonly, buffers are mixtures of a conjugate acid and a conjugate base. An acidic buffer contains weak acid and its salt. An example for this buffer is a mixture of acetic acid (conjugate acid) and sodium acetate (conjugate base), while a basic buffer contains a weak base and its salt. Depending upon the type of buffers, each conjugate acid-base pair has a characteristic pH buffering zone. For instance, phosphate buffer has an effective buffering zone in between 5.9 - 7.9, on the other hand, ammonium ion and ammonia pair have an effective buffering zone between pH 8.3 - 10.3. The importance of buffers and their practical application is explained by the *Henderson-Hasselbalch* equation.

The ionization of a weak acid (HA) and its salt (BA) is represented as:

$$HA \rightleftharpoons H^+ + A^-$$

$$BA \rightleftharpoons B^+ + A^-$$

The dissociation constant K_a for weak acid can be calculated from the following equation:

$$K_a = \frac{[H^+]\,[A^-]}{[HA]}$$

Where, $[H^+]$, $[A^-]$ and $[HA]$ are active concentrations of the compound and its ionic species.

Rewriting the equation,

$$[H^+] = \frac{K_a \times [HA]}{[A^-]}$$

or $\log [H^+] = \log K_a + \log [HA] / [A^-]$, multiplying the equation by -1 yields,

$$- \log [H^+] = -\log K_a - \log [HA] / [A^-]$$

As pH equals $-\log [H^+]$ and pK_a equals $-\log K_a$, hence

$$pH = pK_a + \log \frac{[A^-]}{[HA]}$$

Principally, the A^- ion is derived from ionization of salt (BA), thus *Henderson-Hasselbalch* equation can be rewritten as,

$$pH = pK_a + \log ([salt] / [acid]).$$

A buffer has its maximum buffering capacity at its pK_a (a pH at which the concentration of A^- is equal to HA). Rewriting the *Henderson-Hasselbalch* equation,

$$pH = pK_a + \log 1; \; pH = pK_a + 0$$

$$\text{Thus, } pH = pK_a$$

Normally, a buffer should not be employed at a $pH >1.0$ unit, from its pK_a value. Some of the most commonly used buffers in biochemical analysis include the following:

Buffer	pK_a Value	Buffering pH range
Phosphate	6.86	6.5 - 7.5
Carboxylic acid		
(a) Acetate	4.76	3.0 - 6.0
(b) Citrate	4.74	
Borate	9.24	8.5 - 10.0
Amino acid & peptide		
(a) Glycine	9.6 (pK_{a2})	2.0 - 3.0 & 9.5 - 10.5
(b) Histidine	6.0 (pK_{a2})	5.5 - 6.5
(c) Glycylglycine	8.4 (pK_{a2})	8.0 - 9.0
Zwitterionic* (Good's Buffer)		
(a) Tris	8.10	7.5 - 9.0
(b) MOPS	7.20	6.5 - 7.9
(c) HEPES	7.55	7.0 - 8.0

* Tris - Tri(hydroxymethyl) aminomethane; MOPS- 3-(N-morpholino)-propane sulphonic acid; HEPES - N-2-hydroxyethyl piperazine-N′-2-ethane sulphonic acid.

Calculations for buffer preparations

Example 1: Determine the pH of a buffer solution prepared by mixing 0.075 M acetate and 0.025 M acetic acid. (pK_a of acetic acid = 4.76).

By using the *Henderson-Hasselbalch* equation - $pH = pK_a + \log ([\text{salt}] / [\text{acid}])$

$$pH = pK_a + \log ([\text{acetate}] / [\text{acetic acid}])$$

$$\text{i.e., } pH = 4.76 + \log ([0.075] / [0.025])$$

$$pH = 4.76 + \log 3.0$$

$$pH = 4.76 + 0.4771 = \textbf{5.24}$$

Example 2: Calculate the concentration of acetic acid and acetate in 0.1 M acetate buffer *pH* 5.24. (pK_a of acetic acid = 4.76).

By using the *Henderson-Hasselbalch* equation - $pH = pK_a + \log ([\text{salt}] / [\text{acid}])$

Let x be the molarity of acetate,

then, 0.1 - x = molarity of acetic acid

By substituting these values,

$$5.24 = 4.76 + \log ([x] / [0.1\text{-}x]),$$

$$\text{i.e., } (5.24 - 4.76) = \log ([x] / [0.1\text{-}x])$$

$$0.48 = \log [x] / [0.1\text{-}x],$$

$$\text{i.e., } 3.0 = ([x] / [0.1\text{-}x]), \text{ by taking antilog of } 0.48$$

$$3 (0.1\text{-}x) = x$$

$$0.3 - 3x = x$$

$$\text{i.e., } 0.3 = x + 3x$$

$$0.3 = 4x$$

$$\therefore x = 0.3 \div 4 = 0.075$$

Molarity of acetate is 0.075, then Molarity of acetic acid is
0.1 - 0.075 = 0.025.

Molarity of buffer = **0.075** (salt) + **0.025** (acid) = 0.1 M

Example 3: Calculate the weight of the sodium acetate (in grams) needed to prepare 2 liters of 0.1 M acetate buffer, pH 5.24 and the volume of 1M of acetic acid required to prepare the buffer. (Molecular weight of anhydrous sodium acetate = 82).

By using the *Henderson-Hasselbalch* equation - $pH = pK_a + \log ([salt]/[acid])$

Let x be the molarity of acetate,
then, 0.1 - x = molarity of acetic acid

By substituting these values,
5.24 = 4.76 + log ([x] / [0.1-x]),
i.e., (5.24 - 4.76) = log ([x] / [0.1-x])
0.48 = log [x] / [0.1-x],

i.e., 3.0 = ([x] / [0.1-x]), by taking antilog of 0.48
3 (0.1-x) = x
0.3 - 3x = x
i.e., 0.3 = x + 3x
0.3 = 4x
∴ x = 0.3 ÷ 4 = 0.075

Molarity of acetate is 0.075, then Molarity of acid is 0.1 - 0.075 = 0.025.
Molarity of buffer = **0.075** (salt) + **0.025** (acid) = 0.1 M

Number of moles acetate in 2 L buffer = 2.0 L x 0.075 M = 0.15 moles
Number of moles of acetic acid in 2 L of buffer = 2.0 L x 0.025 = 0.05 moles.

(i)   Weight of sodium acetate in 2 L of buffer = 0.15 moles x 82 = **12.3 g**
(No. of moles = Weight in grams ÷ Molecular weight)

(ii)   Volume of 1 M acetic acid required to prepare 2 L of 0.1 M acetate buffer = 0.05 moles of acetic acid comes from 1 M acetic acid.

$$\text{Number of moles} = \text{Liter x Molarity of the acid}$$
$$0.05 = \text{L x 1 M}$$
$$\text{L} = 0.05 \div 1 = 0.05\ \text{L} = \mathbf{50\ mL}$$

To prepare 2 L of 0.1 M acetate buffer, *p*H 5.24, 12.3 g of sodium acetate and 50 mL of 1 M acetic acid are needed.

(See **appendix** *for preparation of standard buffers for p*H *meter calibration, p*H *indicators for volumetric analysis and p*H *values of frequently used 0.1 M acid & alkaline solutions).*

Additional Reading

1.   Segel, I. H. Biochemical Calculations. 2nd ed. John Wiley & Sons. Inc. New York. (1976).

2.   Beynon, R. J. & Esterby, J. S. Buffer Solutions the Basics. IRL Press, Oxford, (1996).

3.   Nelson, D. L., & Cox, M. M. *Lehninger Principles of Biochemistry*. 3rd ed. Macmillan Press Ltd, UK, Worth Publishers, USA. Reprinted version (2001).

6

Qualitative Analysis
of Amino acids, Proteins, Carbohydrates, Lipids & Steroids

(A) AMINO ACIDS AND PROTEINS

Proteins, the biomacromolecules are polymers of amino acids, held by covalent linkages - *the peptide bonds* and exhibit diverse structures and biological functions. In addition to peptide linkages, the protein structure is stabilized through covalent interactions, namely the disulphide bonds and non-covalent forces such as hydrogen bonding, *van der Waals* forces and electrostatic linkages. Proteins are also polyelectrolytes and the charge is based on the amino acid composition and the amino acid side chains. The four different levels of structural organization in proteins are *primary*, *secondary*, *tertiary* and *quaternary*.

Proteins on hydrolysis yield a mixture of L-α-amino acids. There are 20 naturally occurring amino acids, which vary from one another with respect to their side chains. Amino acids are primary amines, except proline and hydroxy proline which are considered as imino acids. They can be distinguished from rest of the amino acids by ninhydrin reaction, wherein a yellow colour is obtained instead of purple colour. Further, due to the presence of characteristic side chains, certain amino acids exhibit a typical colour reaction, that forms the basis for their identification and determination. Proteins also respond to the characteristic colour reactions of amino acids. However, they can be distinguished from the latter by heat coagulation and biuret reaction.

Traditionally, amino acids of a protein hydrolysate are separated and identified by techniques such as paper, thin layer and ion-exchange chromatography. Amino acids being low molecular weight compounds (average MW of an amino acid residue ~120 Da) when compared to proteins, techniques such as dialysis and gel-filtration are of little value in their separation. Further, the net charge on the amino acids, peptides and proteins is dependent on the pK_a (negative log of the dissociation constant) value of the ionizable groups and the pH of the medium. It should be noted that, at the isoelectric pH (*pI value*), the amino acids, peptides and proteins bear no charge. However, below or above their *pI* value, these molecules have a net positive and negative charge, respectively. This ionic property is exploited in separation of amino acids, peptides and proteins by biochemical techniques such as, ion-exchange chromatography, electrophoresis and isoelectric precipitation.

Aromatic amino acids and proteins exhibit UV absorption. Peptides and proteins containing aromatic amino acids (tryptophan and tyrosine) show UV absorption in the UV-B region (280 - 295 nm) of the electromagnetic spectrum. The absorption at lower wavelengths (in the range of 205 - 220 nm) is due to peptide bonds. However, many other compounds also show absorption in this region.

Qualitative Tests for Amino Acids

Ninhydrin test

Principle: Amino acids react with ninhydrin to yield a purple coloured complex (*Ruhemann's* purple) while, imino acids such as proline and hydroxyproline give yellow colour.

Note: Even, ammonia, primary amines, peptides and proteins react with ninhydrin to give purple to blue colour. When proline reacts with ninhydrin, there is no liberation of ammonia.

(Purple coloured product -
α-amino acid)

(Yellow coloured product –
imino acid)

Reaction Products of Amino Acids with Ninhydrin

Reagent: Ninhydrin reagent: Ninhydrin (2% w/v) in acetone.

Amino acid solution: Prepare 0.1% (w/v) of the individual amino acid (alanine, leucine, lysine and proline) in distilled water.

Procedure: To 1 mL of the amino acid solution taken in a test tube, add few drops of ninhydrin reagent and vortex the contents. Place the test tube in a boiling water bath for 5 min and later cool to room temperature.

Observation: Appearance of purple to blue colour indicates the presence of α-amino acids. In case of imino acids, a yellow colour is observed.

Xanthoproteic test:

Principle: The aromatic groups of either the free amino acids (tyrosine, tryptophan) or proteins, under go nitration to give nitro-derivatives that are yellow in colour in acidic medium. At alkaline *pH*, the colour changes to orange due to the ionization of the phenolic group.

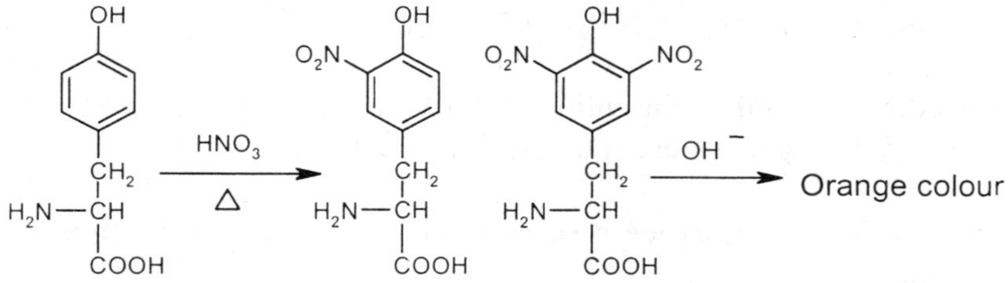

Tyrosine   Mono- or dinitro- derivatives of tyrosine

Reagents: (i) sodium hydroxide (40% w/v); (ii) conc. nitric acid; (iii) hydrochloric acid (6 N).

Amino acid solution: Prepare 0.1%(w/v) of the individual amino acid (tyrosine, tryptophan) in distilled water by adding few drops of 6N hydrochloric acid.

Protein solution: Prepare 0.5%(w/v) solution of bovine serum albumin.

Procedure: To 1 mL of the amino acid or protein solution taken in a test tube, add few drops of nitric acid and vortex the contents. Boil the contents over a Bunsen flame, using a test tube holder, for few minutes. Cool the test tube under running tap water and add few drops of alkali.

Observation: In acid medium the solution is yellow and upon treatment with alkali, the colour changes to orange.

Note: (i) Phenylalanine gives a weak positive reaction for this test (ii) aliphatic amino acids give negative xanthoproteic test (iii) In case of proteins, initially there is precipitation, which goes into solution upon heating.

Ehrlich's test

Principle: The indole ring of tryptophan reacts with *p* - dimethylamino-benzaldehyde under acidic condition to give a purple colour.

Reagent: *p*-dimethylaminobenzaldehyde (10% w/v) in hydrochloric acid.

Tryptophan solution: Prepare 0.1% (w/v) of tryptophan in distilled water by adding few drops of 6 N hydrochloric acid.

Procedure: To 1mL of the amino acid solution taken in a test tube, add few drops of *Ehrlich's* reagent and vortex the contents.

Observation: Appearance of a purple colour indicates the presence of tryptophan.

Note: In case, the colour does not appear, add 1 or 2 drops of conc. hydrochloric acid and mix. Other indole derivatives also respond to this test.

Hopkins- Cole test

Principle: The indole moiety of tryptophan condenses with aldehydes in acidic condition to yield purple or violet coloured compounds.

Reagent: (i) Acetic acid-glyoxylic acid reagent: glacial acetic acid exposed to sun light (for 5 - 6 hours) for the formation of small amounts of glyoxylic acid (ii) conc. sulphuric acid.

Tryptophan solution: Prepare 0.1% (w/v) of tryptophan in distilled water by adding few drops of 6 N hydrochloric acid.

Procedure: Mix 1 mL of the amino acid solution with 1 mL acetic acid-glyoxylic acid reagent, in a test tube, vortex. Then carefully, add conc. H_2SO_4 along the side of the test tube, keeping the tube in an inclined position (do not shake the test tube, while adding the acid).

Observation: A purple - violet ring appears at the junction of the amino acid solution and the conc. sulphuric acid. The formation of this coloured ring confirms the presence of tryptophan.

Lead sulphide test

Principle: Sulphur containing amino acids, such as cysteine and cystine upon boiling with sodium hydroxide (hot alkali) yield sodium sulphide (reaction 1). This is due to partial conversion of the organic sulphur to inorganic sulphide, which can be detected by precipitating it to lead sulphide, using lead acetate solution (reaction 2), as shown below.

Reaction 1:

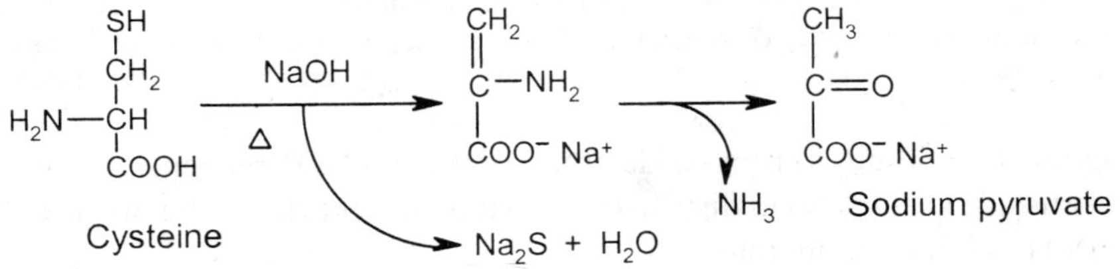

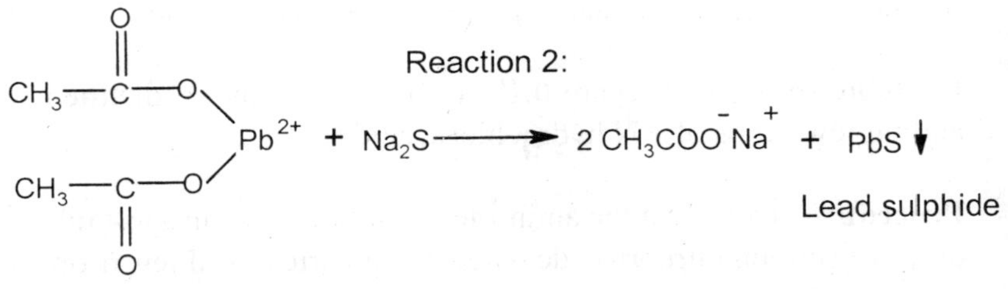

Reaction 2:

Reagents: Lead acetate solution (10% w/v); Sodium hydroxide solution (40% w/v) solution.

Cysteine solution: Prepare 0.1% (w/v) of cysteine in distilled water by adding few drops of 1 N hydrochloric acid.

Procedure: To 1mL of the amino acid solution taken in a test tube, add few drops of sodium hydroxide (40%) and boil the contents for 5-10 min over a Bunsen burner. Cool the contents and add few drops of 10% lead acetate solution and observe.

Observation: The appearance of a black precipitate of lead sulphide confirms the presence of cysteine or cystine.

Note: The sulphur present in methionine is not released by alkali treatment, thus gives a negative test.

Sodium nitroprusside test

Principle: Sodium nitroprusside reacts with the thiol group of the cysteine under alkaline condition to yield an intense purple coloured compound, which fades upon standing.

Note: Cystine gives a positive test only after reduction of the disulphide group to sulphydryl (thiol) group, by chemical agents such as potassium or sodium cyanide. This test is not specific for mercaptans, since compounds such as acetone, acetoacetate and creatine also respond to this test.

Reagent: (i) Sodium nitroprusside $\{[Na_2Fe (CN)_5] NO\}$ reagent: sodium nitroprusside (10% w/v) in distilled water (this reagent should be prepared freshly) (ii) liquor ammonia.

[**Caution**: (i) use chilled ammonia (ii) use pro-pipette for dispensing].

Cysteine solution: Prepare 0.1% (w/v) of cysteine in distilled water by adding few drops of 1 N hydrochloric acid.

Procedure: To 1mL of the amino acid solution taken in a test tube, add few drops of sodium nitroprusside reagent and vortex. Add few drops of liquor ammonia mix the contents.

Observation: An intense purple colour indicates the presence of cysteine. The intensity of the colour fades with time.

Sullivan and McCarthy's test

Principle: Addition of sodium nitroprusside to an alkaline solution of methionine followed by acidification of the reaction yields a red colour. This reaction also forms the basis for the quantitative determination of methionine.

Note: The amino acids tryptophan and histidine interfere in this test.

Reagents: (i) Sodium hydroxide (5 N) solution (ii) Glycine (2% w/v) solution (iii) Sodium nitroprusside (10% w/v) (iv) 6 N HCl.

Methionine solution: Prepare 0.1% (w/v) in distilled water.

Procedure: To 1 mL of the amino acid solution taken in a test tube, add few drops of sodium hydroxide (5 N), followed by addition of few drops of glycine (2%) and 10% sodium nitroprusside solution and vortex. Place the test tube in a hot water bath, maintained at 40°C, for 15 minutes. Cool the tube in ice cold water for 5 minutes and add 0.5 mL of 6 N HCl. Vortex the contents and allow to stand for15 min at room temperature.

Observation: A strong red colour indicates the presence of methionine.

Sakaguchi test

Principle: Under alkaline condition, α-naphthol (1-hydroxy naphthalene) reacts with a mono-substituted guanidine compound like arginine, which upon treatment with hypobromite or hypochlorite, produces a characteristic red colour. The reaction mechanism is poorly understood.

Note: In addition to arginine, compounds such as α-guanidino butyric acid, mono-methylguanidine and octopine, tests positive to this reaction, while guanidine, dimethylguanidine, nitro-arginine, cirtulline, creatinine and creatine do not respond to this reaction.

Reagents: (i) Sodium hydroxide (10% w/v) (ii) α-naphthol reagent (1% w/v in ethyl alcohol) (iii) Hypobromite reagent (to be freshly prepared): Take 100 of 5% (w/v) sodium hydroxide solution in a glass reagent bottle and add 1 mL of pre-chilled liquid bromine, using a pro-pipette. Shake the contents till bromine dissolves (iv) Urea solution 5% (w/v).
Caution: Use chemical safety hood, while adding bromine (corrosive & toxic chemical).

Amino acid solution: Prepare 0.1% (w/v) of arginine in distilled water.

Procedure: To 1mL of chilled amino acid solution taken in a test tube, add few drops of sodium hydroxide solution, followed by α-naphthol reagent, and vortex. After 2 min add few drops of 5% urea solution, followed by freshly prepared hypobromite reagent, drop wise, until the desired colour is achieved.

Observation: Formation of a red colour denotes the presence of arginine.

Isatin test

Principle: Imino acids such as proline and hydroxy proline condense with isatin under acidic conditions to yield a blue coloured adduct.

Isatin

Reagent: Isatin reagent: isatin (1% w/v) in acetic acid.

Imino acid solution: Prepare 0.2% (w/v) of proline or hydroxy-proline in distilled water.

Procedure: Apply a drop of imino acid solution on a Whatman No. 1 filter paper strip (size 25 x 50 mm) and dry the spot using a hot air gun / hair dryer or in a hot air oven. Apply a drop of isatin reagent on to the dried spot. Repeat the drying procedure with hot air gun for few minutes and observe.

Observation: Appearance of a characteristic blue coloured spot on the filter paper, confirms the presence of imino acid.

Note: Other amino acids give pink colour with this reagent.

Pauly's diazo test

Principle: Sulphanilic acid upon diazotization in the presence of sodium nitrite and hydrochloric acid results in the formation a diazonium salt. The diazonium salt formed, couples with either tyrosine or histidine in alkaline medium to give a red coloured chromogen (azo dye).

Diazotization and coupling reaction

Reagents: (i) sulphanilic acid (1% w/v) in 1 N hydrochloric acid (ii) sodium nitrite (5% w/v) in distilled water (to be freshly prepared) (iii) sodium carbonate (10 % w/v) in distilled water.

Amino acid solution: Individually, prepare 0.1% (w/v) of tyrosine and histidine in distilled water by adding few drops of 6 N hydrochloric acid.

Procedure: Take 1mL of sulphanilic acid reagent in a test tube and chill the contents in a small ice bucket. Add few drops of pre-chilled sodium nitrite solution and vortex. Add immediately few drops of pre-chilled amino acid solution and vortex. This is followed by drop wise addition of sodium carbonate solution until the colour appears.

Observation: Appearance of a red colour denotes the presence of either tyrosine or histidine.

Millon's test

Principle: The phenolic group of tyrosine reacts with mercuric ions in acidic condition in the presence of sodium nitrite, yielding a red colour complex.
Note: Chlorides, phosphates interfere in the colour development. Tryptophan yields a reddish-brown colour with this reagent.

Reagents: (i) *Millon's* reagent (15% w/v mercuric sulphate in 6 N sulphuric acid) (ii) Sodium nitrite solution (5%) – to be freshly prepared.

Amino acid solution: Prepare 0.1% (w/v) of tyrosine in distilled water by adding few drops of 6 N hydrochloric acid.

Procedure: To 1 mL of the amino acid solution in a test tube, add few drops of *Millon's* reagent and vortex. Boil the contents over a Bunsen flame for 3 - 5 min. Cool the contents under running tap water and add few drops of sodium nitrite solution.

Observation: Formation of a red colour confirms the presence of tyrosine.
Note: Histidine gives a negative *Millon's* test.

Gerngross test

Principle: Nitrosonaphthol reacts with tyrosine in acid medium to give a purple-red coloured complex.

Reagents: (i) Nitrosonaphthol reagent: nitrosonaphthol (0.1% w/v) in ethanol (ii) conc. nitric acid.

Amino acid solution: Prepare 0.1% (w/v) of tyrosine in distilled water by

Flow sheet: Qualitative Analysis of Amino acids

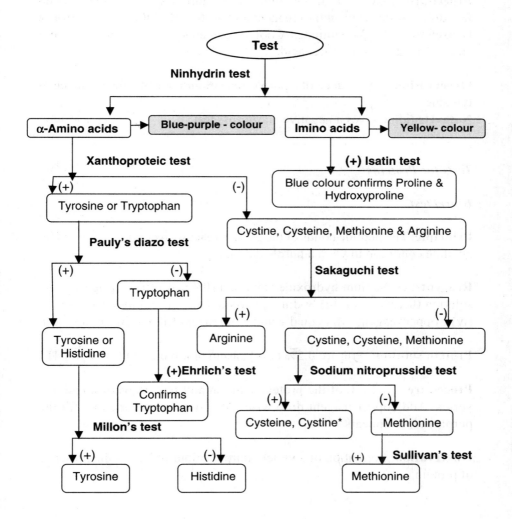

* Cystine shows a positive nitroprusside test after reduction of the disulphide bonds with reducing agents like KCN, or NaCN.

Note: This flow sheet is applicable for qualitative analysis of eight amino acids. However, the rest of the amino acids which show positive ninhydrin reaction can be identified by paper, TLC and ion-exchange chromatography.

adding few drops of 6 N hydrochloric acid.

Procedure: Take 1 mL of the amino acid solution in a test tube and add few drops of nitrosonaphthol reagent and vortex. Boil the contents over a Bunsen flame for 3 - 5 minutes. Cool the contents under running tap water and add 1 - 2 drops of nitric acid.

Observation: Appearance of a purple-red colour confirms the presence of tyrosine.
Note: Histidine gives a negative *Gerngross* test.

Tests for Proteins

Biuret test

Principle: The peptide bonds of the protein react with cupric copper, under alkaline condition to yield a purple colour.

Reagents: (i) Sodium hydroxide solution (10% w/v) (ii) Cupric sulphate solution (0.15% w/v) (iii) Sodium potassium tartrate solution (0.6% w/v) (iv) Copper reagent: Mix equal volume of reagent (ii) and (iii) before test.

Protein solution: Prepare 0.5% (w/v) casein or albumin in 0.1 N NaOH.

Procedure: To 2 mL of the protein solution add few drops of alkali and vortex. Add copper reagent drop wise with continuous vortexing till the purple colour appears.

Observation: Formation of a violet - purple colour indicates the presence of protein.

Ammonium sulphate precipitation test

Principle: Globular proteins get precipitated at half saturation (50%) with ammonium sulphate, while albumins precipitate at full saturation of the salt. This salting out behaviour is attributable to weakening of the interactions between protein and its bound water (hydrated) in the presence of salt. This is followed by increased protein-protein interactions, leading to aggregation or precipitation.

Reagent: (i) Saturated ammonium sulphate solution (ii) solid ammonium sulphate.

Material: Hen's egg.

Procedure: Carefully break the shell of an egg and collect the egg white in a 100 mL beaker. Make 1: 1 dilution of the collected egg white with distilled water. Take 2 mL of the diluted protein solution in a glass centrifuge tube (15mL capacity) and add equal volume of saturated ammonium sulphate solution with gentle stirring (avoid frothing). Centrifuge (2500 - 3000 rpm for 10 min) to separate the precipitate formed, if any. Decant the supernatant carefully in a test tube. Saturate this supernatant by addition of solid ammonium sulphate with constant stirring.

Observation: The precipitate observed at 50 and 100% saturation of ammonium sulphate is that of globulins and albumins, respectively.

Sulphosalicylic acid test

Principle: Sulphosalicylic acid precipitates proteins from solutions. This is due to the interaction of sulphosalicylic acid with the basic groups of the protein, resulting in protein aggregation and precipitation.

Reagent: Sulphosalicylic acid reagent: Prepare, sulphosalicylic acid 5% (w/v) in 20% (w/v) sodium sulphate solution.

Material: Bovine serum albumin solution (0.5-1.0% w/v).

Procedure: To 2 mL of the sample solution add sulphosalicylic acid reagent drop wise with constant mixing.

Observation: Appearance of a white precipitate or turbidity in the sample solution indicates the presence of protein(s).

Heat coagulation test

Principle: Proteins in acidic medium, upon heating undergo thermal denaturation, leading to their aggregation and precipitation.

Protein solution: Egg white solution.

Reagent: Acetic acid (1% v/v).

Procedure: Take 2 mL of the protein or sample solution in a test tube. Acidify the protein solution by addition of few drops of acetic acid followed by heating the contents over a Bunsen flame for few minutes.

Observation: The appearance of a precipitation (coagulum) indicates presence of protein(s).

(B) CARBOHYDRATES

Carbohydrates are polyhydroxy aldehydes or ketones that occur in nature as monosaccharides, disaccharides, oligosaccharides and polysaccharides. The monosaccharides and disaccharides exhibit reducing property owing to the presence of a free hydroxyl group at the anomeric carbon atom. Hence, reducing sugars show mutarotation, undergo oxidation and react with phenylhydrazine to form osazones. However, this reducing property is lacking in certain disaccharides (ex. sucrose, trehalose) and polysaccharides (ex. starch, glycogen).

Monosaccharides undergo dehydration under acidic condition to furfural or hydroxymethyl furfural. Formation of these aldehydes forms the basis for certain qualitative and quantitative reactions of sugars. In addition, oxidation of sugar to sugar acids is also a characteristic property of carbohydrate reactions, which forms the basis for qualitative and quantitative analysis, as well.

Qualitative tests for carbohydrates

Molisch test: This is a general test for the identification of carbohydrates.

Principle: In the presence of acid, the sugars undergo dehydration to furfural or hydroxymethyl furfural, which condenses with α-naphthol (1-hydroxy naphthalene), resulting in the formation of a purple coloured ring. The acid (H_2SO_4) employed in this reaction not only catalyses the dehydration of sugars, but also brings about the hydrolysis of glycosidic bonds of oligo- and polysaccharides.

Reagents: (i) α-naphthol reagent (5% w/v in ethyl alcohol) (ii) Conc. Sulphuric acid.

Sugar solution [1% (w/v) in distilled water]: Prepare individual sugar solutions of glucose, ribose, lactose and starch.

Procedure: To 1mL of the sugar solution taken in a test tube, add few drops of α-naphthol reagent and vortex the contents. Carefully, add conc. H_2SO_4 along the side of the test tube, keeping the tube in an inclined position (do not shake the test tube, while adding the acid).

Observation: A purple ring appears at the interphase of sugar solution and the conc. sulphuric acid. The formation of this purple coloured ring confirms the presence of carbohydrates.

Iodine test:

Principle: Molecular iodine reacts with polysaccharide micelles to form a complex that are ill defined. Variations in the colour of the polysaccharide-iodine complex are observed (starch gives blue colour with iodine, while glycogen forms a red-brown coloured complex).

Reagent: Prepare a 3% (w/v) solution of potassium iodide in distilled water. Add few crystals of iodine until the solution becomes deep yellow in colour.

Sugar solution: Prepare 1% (w/v) of polysaccharide (starch / glycogen) in distilled water.

Procedure: To 2 mL of the sugar / test solution add few drops of iodine reagent and vortex.

Observation: Appearance of a blue colour indicates the presence of starch, whereas a reddish-brown colour is indicative of glycogen. Interestingly, the colour of the solution disappears on heating and reappears on cooling.

Test for reducing sugars:

Principle: The cupric (Cu^{2+}) copper present in the alkaline copper sulphate solution is reduced to cuprous (Cu^+) hydroxide by the reducing sugar (reaction 1). However, this metal hydroxide is unstable and undergoes spontaneous dehydration to cupric oxide (reaction 2), which appears as a coloured precipitate (red, brick-red, orange, green). In this reaction, the reducing sugar is oxidized to sugar acid (aldonic acid).

Reaction 1

Sugar   Sugar acid

Reaction 2

$$2 \left[Cu^+ \ OH^- \right] \xrightarrow{\quad\quad} Cu_2O \downarrow$$

Cuprous hydroxide (unstable) H_2O Cupric oxide

This principle forms the basis for the following biochemical tests (i) *Fehling's* (ii) *Benedict's*, and (iii) *Barfoed's*. However in *Barfoed's* test, the reduction of cupric ions is under mild acidic condition and the test is more rapid for monosaccharides than disaccharides.

Fehling's test:

Reagents: Reagent A : Dissolve 6.92 g of cupric sulphate in distilled water and make up the volume to 100 mL in a volumetric flask. Store this reagent in a plastic reagent bottle. Reagent B: Dissolve 34.6 g of sodium potassium tartrate and 25 g of potassium hydroxide in distilled water and make up the volume to 100 mL with distilled water and store in a reagent bottle.

Sugar solution [1 %(w/v) solution in distilled water]: Prepare individual sugar solutions of glucose, fructose, xylose, sucrose, lactose.

Procedure: Mix 1 mL each of reagents A and B in a test tube. Add few drops of sugar solution/ test sample and vortex. Heat the tube in a boiling water bath for 10 min.

Observation: Appearance of a coloured precipitate (orange to brick red) indicates the presence of a reducing sugar. However, this test is not reliable, as compounds other than reducing sugars also test positive to this reaction (ex. Chloroform, ammonium salts).

Benedict's test

Reagents: Dissolve, 17.3 g of sodium citrate and 10 g of sodium carbonate in 50 mL of distilled water and warm the solution in a hot water bath, with constant stirring. Separately, prepare cupric sulphate solution by dissolving 1.73 g of this salt in 20 mL of distilled water. Mix the cupric sulphate solution with sodium citrate -carbonate solution and make up the volume to 100 mL. Store the reagent in a reagent bottle.

Sugar solution [1 %(w/v) solution in distilled water]: Prepare individual sugar solutions of glucose, fructose, xylose, sucrose, lactose.

Procedure: To 2 mL of *Benedict's* reagent taken in a test tube, add few drops of sugar solution / test sample and place the tube in a boiling water bath for 5 min.

Observation: Appearance of a coloured precipitate (green/ yellow/ orange/ red) confirms the presence of reducing sugar.

Barfoed's test

Reagent: Dissolve, 13.3 g of cupric acetate in 80 mL of distilled water. Add 2 mL of glacial acetic acid and make up the volume to 100 mL with distilled water.

Sugar solution [1 %(w/v) glucose or maltose solution in distilled water].

Procedure: To 2 mL of the reagent taken in a test tube, add few drops of the sugar / test solution and heat the contents in a boiling water bath.

Observation: The appearance of a red precipitate as a thin film at the bottom of the test tube with in 3 - 5 minutes is indicative of a reducing monosaccharide. If the time taken for the formation of the precipitate is more, it is suggestive of a reducing disaccharide.

Note: Anomeric hydroxyl group of sugar blocked by either alkylation (α- or β- methyl glycosides) or by glycosidic bond formation with other monosaccharide (ex. sucrose) do not respond to these tests . (ii) In case of reducing disaccharides (ex. maltose and lactose), one of the anomeric hydroxyl group of the disaccharide is not involved in the glycosidic bond formation, hence shows positive to reducing tests.

Seliwanoff's test

Principle: Ketosugars (ex. fructose) undergo dehydration in acid medium to hydroxymethyl furfural more rapidly than aldohexoses. This derivative formed condenses with resorcinol (*m*-dihydroxy benzene) to yield a pink colour.

Reagent: (i) Prepare 0.05% (w/v) resorcinol in 4 N hydrochloric acid.

Sugar solution [1 %(w/v) fructose solution in distilled water].

Procedure: To 2 mL of the *Seliwanoff's* reagent taken in a test tube, add few drops of sugar / test solution and place the tube in a boiling water bath for 5 minutes.

Observation: Appearance of a pink colour confirms the presence of a keto sugar.

Note: High concentration of glucose or other sugars may interfere by producing similar coloured compounds with *Seliwanoff's* reagent.

Bial's test:

Principle: Pentose sugars undergo dehydration in acid medium to furfural, which condenses with orcinol (3, 5-dihydroxy toluene) to yield a green coloured complex.

Reagent: Prepare 0.3% orcinol in concentrated hydrochloric acid and add 0.25 mL of 10% (w/v) ferric chloride solution.

Sugar solution: [1 %(w/v) ribose/xylose solution in distilled water].

Procedure: To 2 mL of the *Bial's* reagent, taken in a test tube, add few drops of the sugar/ test solution and heat in a boiling water bath for 5-10 minutes.

Observation: Appearance of a green colour is indicative of the presence of pentose sugar.

Mucic acid test:

Principle: Monosaccharides, upon treating with strong oxidizing agents, such as nitric acid, yield saccharic acids. The saccharic acid obtained after oxidation of galactose is insoluble and separates as gritty crystals. This acid derivative is known as mucic acid, thus the name for the test. Galactose containing saccharides tests positive to this reaction.

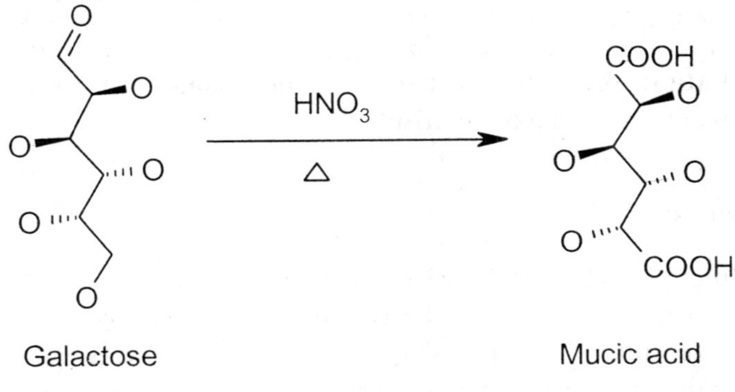

Galactose   Mucic acid

Reagent: Concentrated nitric acid.

Sugar solution: Galactose (2% w/v) in distilled water.

Procedure: Take 5 mL of the galactose solution in a 50 mL glass beaker and add 2 mL of conc. nitric acid. Concentrate the contents to a small volume (2 - 3 mL) by heating (over a flame or on steam bath). After concentration, cool the solution gradually to room temperature.

Observation: Appearance of gritty crystals of mucic acid confirms the presence of galactose.

Test for sucrose

Principle: Sucrose, the nonreducing disaccharide on acid hydrolysis forms a mixture of reducing sugars (glucose and fructose), that can be confirmed by *Benedict's* and *Seliwanoff's* test.

Reagents: (i) Conc. hydrochloric acid (ii) *Benedict's* reagent (iii) *Seliwanoff's* reagent (iv) Sodium hydroxide (10 N).

Sugar solution: Sucrose solution (1 % w/v) in distilled water.

Procedure: Take 3 mL of the sugar solution in a test tube and add few drops of conc. HCl and heat the tube in a boiling water bath for 20 min. Cool the solution to room temperature and test the solution for the presence of reducing and ketosugar by *Benedict's* and *Seliwanoff's* reagents.

Note: Before testing the solution by *Benedict's* reagent, neutralize the acid hydrolysate by adding few drops of 10N NaOH.

Observation: A positive test for *Benedict's* and *Seliwanoff's* reaction, confirms the presence of sucrose.

Osazone test

Principle: Phenylhydrazine reacts with reducing sugars to yield osazones, which are the characteristic derivatives of carbohydrates. Each osazone crystallizes into a definite shape. Thus, it is possible to identify the carbohydrate by this test. Non-reducing sugars do not react with phenylhydrazine as they lack an anomeric hydroxyl group. Glucose, fructose

Osazone reaction – formation of glucosazone

and mannose, upon reaction with phenylhydrazine yield only one type of osazone, i.e., glucosazone. Each sugar osazone differs from the other with respect to time of crystallization, crystal shape and melting point.

Sugar solution: Prepare 2% (w/v) sugar (glucose, fructose, galactose, xylose, lactose and maltose) solution in distilled water.

Reagents: (i) Phenylhydrazine reagent: This reagent is prepared by mixing and grinding, 2 parts of phenylhydrazine hydrochloride and 3 parts of sodium acetate (or equal part of anhydrous sodium acetate) in a glass mortar & pestle (ii) glacial acetic acid.

Equipment: Light microscope.

Procedure: To 3 mL of the sugar solution taken in a test tube, add 0.5 g of phenylhydrazine reagent and few drops of acetic acid. The contents are vortexed and placed in a boiling wate bath for 15 minutes. Cool the solution to room temperature and observe the shape of the crystals formed, under a light microscope.

Observation: Microscopic examination of the osazones reveals the following shapes of crystals:

(i) glucose, mannose and fructose - *needle shaped crystals arranged singly or in groups (Feathery).*

(ii) xylose   - *long fine needles.*

(iii) maltose   - *sunflower shaped crystals.*

(iv) lactose   - *puff shaped crystals.*

Shapes of sugar-osazone crystals

Xylosazone   Maltosazone   Lactosazone

Glucosazone

Time taken for the formation of osazone of different sugars

Sugar	Time (min)
Glucose	5.0
Fructose	2.0
Mannose*	1.0 – 5.0
Galactose	20.0
Xylose	7.0
Arabinose	10.0

* Note: Mannose rapidly forms an insoluble white phenylhydrazone at room temperature, that forms an osazone on heating.

Flow sheet: Qualitative Analysis of Carbohydrates

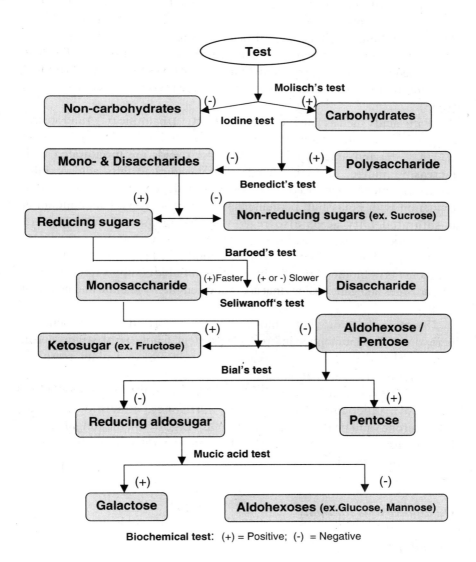

Biochemical test: (+) = Positive; (-) = Negative

(C) LIPIDS & STEROIDS

Lipids, the hydrophobic biomolecules are insoluble in water, but are soluble in organic solvents such as chloroform, benzene, diethyl ether, hexane etc. The simple lipids such as, neutral fats (triglycerides) on hydrolysis yield glycerol and three molecules of fatty acids. If the hydrolysis of fats is performed in the presence of alkali (NaOH. or KOH), the process is termed as *saponification*, as the fatty acids liberated are present as their alkali salts (soap).

Fatty acids such as lauric (C_{12}), myristic (C_{14}), palmitic (C_{16}) and stearic (C_{18}) acids are saturated, while oleic ($C_{18} \Delta^9$), linoleic ($C_{18} \Delta^{9, 12}$), linolenic ($C_{18} \Delta^{9, 12, 15}$) and arachidonic ($C_{20} \Delta^{5, 8, 11, 14}$) acids are unsaturated with increasing number of double bonds. The unsaturated fatty acids exhibit typical chemical properties of olefins such as, halogenation, hydrogenation, oxidation with potassium permanganate and ozonolysis. Fats containing unsaturated fatty acids also react with halogens and the extent of halogenation is a measure of degree of unsaturation.

Phospholipids are also classified as saponifiable lipids and are important components of biomembranes. The non-saponifiable complex lipids include cerebrosides and gangliosides that are found commonly in brain and neuronal tissues.

Steroids, the hydrophobic molecules are categorized by the presence of perhydrocyclopentanophenanthrene ring system. Cholesterol, stigmasterol, lanosterol and ergosterol are examples of animal and plant steroids. Cholesterol, the principal animal sterol is the precursor for a number of bioactive steroids such as androgens (testosterone), estrogens (estrone, estradiol), glucocorticoids (cortisol, cortisone), mineralocorticoids (aldosterone) and progestins (progesterone). Ergosterol, the fungal steroid is a precursor for vitamin D_3. These steroid derivatives exhibit diverse physiological functions.

Qualitative tests for lipids

Solubility test

Principle: Fats and fatty acids (long chain) are insoluble in water as they are hydrophobic in nature. However, they are soluble in organic solvents.

Solvents: Chloroform, ether, benzene, carbon tetrachloride, hexane.

Test samples: Fats -Tristearin, vanaspathi (hydrogenated fat); Fatty acids - palmitic, stearic and oleic acid.

Procedure: Take small quantities of fat / fatty acid into separate test tubes and test their solubility in the above listed organic solvents (3 mL/test tube) and distilled water.

Observation: Palmitic acid, stearic acid, tristearin (solids) do not dissolve in water, while oleic acid (liquid) separates into two phases. All the test samples dissolve clearly in organic solvents.

Saponification test

Principle: Fats on alkaline hydrolysis yield glycerol and the Na^+ or K^+ salt of fatty acid (soap).

Reagent: Alcoholic-KOH (2% w/v KOH in ethyl alcohol).

Test sample: Tristearin, coconut oil.

Procedure: To a small quantity of fat or oil (~100 mg) taken in a test tube, add 3 mL of alcoholic-KOH and vortex. Place the tube in a boiling water bath for 15 - 20 min.

Observation: The test sample goes into solution, which upon shaking results in foaming due to the formation of soap.

Acrolein test

Principle: Glycerol or glycerol containing compounds on heating with potassium bisulphate undergo dehydration to yield acrylic aldehyde or acrolein, which has a characteristic pungent odour.

$$\begin{array}{c}
H_2C-OH \\
| \\
HC-OH \\
| \\
H_2C-OH
\end{array}
\quad \xrightarrow[\Delta]{KHSO_4}
\quad
\begin{array}{c}
H\diagdown C\diagup\!\!\!\!=O \\
| \\
CH \\
\| \\
CH_2
\end{array}$$

$$- 2H_2O$$

Glycerol   Acrylic aldehyde or Acrolein

Reagent: Potassium bisulphate (solid).

Test sample: Glycerol, tristearin, groundnut oil.

Procedure: To a small quantity of the test sample, taken in a test tube, add solid potassium bisulphate (~100 mg) and heat the contents over a Bunsen flame for few minutes.

Observation: Detection of a pungent smell confirms the presence of glycerol.
Note: Free fatty acids give a negative acrolein test.

Tests for unsaturation

Principle:

(i) The double bonds present in the unsaturated fats and fatty acids undergo halogenation (bromination or iodination) to yield di-halo adducts. The extent of decolourization is a measure of degree of unsaturation.

(ii) Unsaturated fats or fatty acids undergo incomplete oxidation to diols upon treatment with alkaline potassium permanganate solution, which results in decolourization of potassium permanganate.

$$\begin{array}{c}
\qquad\qquad X \\
\qquad\qquad | \\
\xrightarrow{X_2} \quad ---HC-CH--- \\
\qquad\qquad\qquad | \\
\qquad\qquad\qquad X
\end{array}$$

Dihalo adduct

$$---\overset{\displaystyle |}{\underset{\displaystyle |}{C}}=\overset{\displaystyle |}{\underset{\displaystyle |}{C}}---$$

Unsaturated fatty acid

$$\xrightarrow[KMnO_4]{OH^-} \quad
\begin{array}{c}
\qquad OH \\
\qquad | \\
---HC-CH--- \\
\qquad\qquad | \\
\qquad\qquad OH
\end{array}$$

Diol derivative

Reagents: (i) Bromine water (ii) dilute potassium permanganate solution (0.05% w/v) (iii) Fat solvent - chloroform.

Test sample: Oleic acid, vanaspathi (hydrogenated fat), coconut oil.

Procedure: Take small quantity of the test sample dissolved in 2 mL chloroform, into a test tube. Add bromine water or dilute potassium permanganate solution drop wise into the test tube with vortexing.

Observation: Decolourization of bromine water or dilute potassium permanganate solution is observed for unsaturated fats or fatty acids, while no decolourization occurs in saturated fats and fatty acids.
Note: Excess reagent addition should be avoided.

Hubl's iodine test

Reagent: (i) *Hubl's* iodine solution: Separately, dissolve 2.6 g of iodine and 6.0 g of mercuric chloride in 40 mL of ethanol (95% v/v). Mix the both and make up the volume to 100 mL with ethanol (95% v/v) (ii) Solvent: chloroform.

Test sample: Oleic acid, vanaspathi (hydrogenated fat), sunflower oil.

Procedure: Take small quantity of the test sample dissolved in 2 mL chloroform, into a test tube. Add *Hubl's* reagent drop wise while vortexing the contents.

Observation: Decolourization of iodine is observed, confirming the presence of unsaturated fats or fatty acids. However, further addition of the reagent results in retention of the colour, indicating the completion of iodination reaction.

Test for steroids

Principle: Steroids in the presence of an anhydride (like acetic anhydride) and conc. sulphuric acid develop colours ranging from blue to green.

Although, the exact reaction mechanism is unknown, it is presumed that the steroid undergoes acid catalysed dehydration to form short lived transitory chromophoric intermediates which arise by rearrangement of double bonds.

*Note: The formation of green chromophore, is preceeded by the formation of various coloured intermediates (orange - red - reddish-purple - blue-green to green).

Salkowski test

Test sample: Cholesterol (0.5% w/v) in chloroform.

Procedure: To 2 mL of the test sample, taken in a test tube, add an equal volume of conc. sulphuric acid along the sides of the tube. Observe the colour in acid and chloroform phase.

Observation: The acid layer shows a yellow colour with green fluorescence, while the chloroform layer is cherry red.

Liebermann-Burchard test

Reagents: (i) Acetic anhydride (ii) Conc. sulphuric acid.

Test sample: Cholesterol (0.5% w/v) in chloroform.

Procedure: To 2 mL of the test sample, taken in a test tube, add an equal volume of acetic anhydride, followed by addition of 2 - 3 drops of conc. sulphuric acid. Vortex the contents.

Observation: Development of a green colour confirms the presence of cholesterol.

Flow sheet: Qualitative Analysis of Lipids

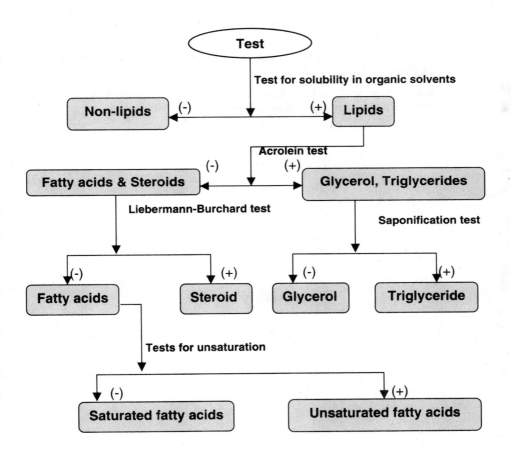

Biochemical test: (+) = Positive; (-) = Negative

Additional Reading

1. Greenstein, J. P. & Winitz, M. Chemistry of the Amino acids. Vol. III. John Wiley & Sons Inc. NY, USA, (1961).

2. Meister, A. Biochemistry of Amino acids. Vol. I, 2nd ed. Academic Press Inc. NY, USA, (1965).

3. Hawk, P.B. Oser, B L. & Summerson, W. H. Practical Physiological Chemistry. 14th ed. Churchill, London, (1966).

4. Kleiner, I. S. & Dotti, L. B. Laboratory Instructions in Biochemistry. The C.V. Mosby Company, St. Louis, USA, (1966).

7

Quantitative Analysis

The realm of quantitative biochemical analysis involves determination of specific analyte(s) in the samples of biological origin. Quantitation of biologically important organic and inorganic analytes by titrimetry and photometry, are based on the established physico-chemical principles. These methods are routinely used to assess the levels or concentration of analytes in biological samples. These methods find wide applications in analytical biochemistry.

One of the conventionally employed method of biochemical analysis is titrimetry. This method is popular because of its simplicity and inexpensive inputs. In the titrimetric analysis, the analyte of interest is allowed to react with an appropriate chemical reagent, which is added as a standardized solution. The completion of the chemical reaction is ascertained by measuring the volume of the standardized solution consumed. The use of indicator indicates the equivalence point of titration by change in colour, which is also indicative of the end point of the reaction. The quantitation is based on the *stoichiometric* calculations. Some of the common types of reactions used in biochemical analysis, in titrimetry include (i) acid-base neutralization (ii) oxidation-reduction (iii) complex formation and (iv) precipitation reaction.

Photometric & UV-Vis spectrophotometric analyses are methods of choice in the analytical biochemical investigations, due to their high sensitivity, specificity, repeatability and reproducibility. The interaction of biomolecules or biomolecule derivatives with radiant energy, both in UV & visible regions of the electromagnetic spectrum forms the basis for spectrophotometric analysis. Measurement of specific absorbance at fixed wavelengths are most frequently used photometric method to obtain quantitative information, such as, concentration of biomolecules or their chromophores [analytes in solution upon reaction with specific reagent(s), yield coloured derivatives] in solution. For detailed information see, *4 - Basic concepts and use of instruments in biochemical analysis*.

Experiments on (A) titrimetric and (B) photometric analysis of a variety of analytes are described in this chapter.

(A) TITRIMETRIC EXPERIMENTS

Experiment #1: Estimation of ascorbic acid [§]

Principle: Ascorbic acid (vitamin C), reduces 2,6-dichlorophenol indophenol, a coloured dye, to colourless leuco form in acidic medium. The vitamin gets oxidised to dehydroascorbic acid. Though the dye is blue in colour, the appearance of a pale pink colour indicates the end point of titration.

L- Ascorbic acid	2,6-dichlorophenol indophenol	Dehydro-ascorbic acid	Leuco dye
	(Blue dye)		(Colourless - reduced dye)

Standard ascorbic acid stock solution (1 mg/mL) : (i) Weigh 100 mg of ascorbic acid transfer it carefully into a 100 mL volumetric flask. Dissolve the vitamin and make up the volume to 100 mL with oxalic acid solution (4%). (ii) Working standard (0.1 mg/mL): Dilute 10 mL of the stock solution to 100 mL, with 4% oxalic acid in a 100 mL volumetric flask.

Reagents: (i) Oxalic acid solution 4% (w/v) (ii) Dye solution: Weigh 26 mg of the dye and 21 mg of sodium bicarbonate and transfer them into a 100 mL volumetric flask. Dissolve and make up the volume to 100 mL with distilled water. Filter the reagent, before use.

Procedure

Pipette out 5 mL of the working standard solution of ascorbic acid into a 100 mL conical flask. Add 10 mL of oxalic acid solution and titrate the contents against the dye solution, taken in a glass burette. The appearance of a pale pink colour, which persists for few minutes, indicates the end point of the titration. There is no need for addition of an indicator, as this dye is a self indicator. Repeat the titration thrice. The volume of dye consumed is equivalent to the amount of vitamin C.

Calculation: Ascorbic acid content (mg/100g sample) is calculated by the following equation;

$$\text{Vitamin C (mg/100 g)} = \frac{0.5 \text{ mg}}{V_2 \text{ mL}} \times \frac{V_1 \text{ mL}}{5 \text{ mL}} \times \frac{100 \text{ mL} \times 10}{\text{Weight of the sample (g)}} \times 100$$

Where,

V_1 = Volume of the dye consumed for standard ascorbic acid (mL).
V_2 = Volume of dye consumed for the sample (mL).
x 10   = Dilution factor.

Workout : Determine the vitamin C content in the following citrus fruits, (i) Orange (ii) Sweet lemon (iii) Lemon. Report the vitamin C content per 100 g of the fruit. Which is the vitamin C rich fruit, among the citrus fruits analysed?

Procedure: Weigh 10 g of citrus fruit and extract the juice into a 100 mL beaker. Filter the extract / juice through a glass funnel plugged with glass wool (soak the glass wool in water before use) into a 100 mL volumetric flask. Make up the volume to 100 mL with 4% oxalic acid. Make a 1:10 dilution of this solution with 4% oxalic acid, before titration. Use 5 mL sample volume for titration. Rest of the procedure is similar to standard titration.

Caution: Do not handle dry glass wool with bare hands, use disposable plastic gloves.

[§](i) Harris, L. J. & Ray, S. N., Lancet, i , 71: 462.(1935).
 (ii) Analytical Uses of 2, 6-DCIP. Product Information Bulletin, J. T. Baker Chemical Company, Commodity # H114, New Jersey, USA (1971).

Experiment #2: Determination of calcium [§]

Principle: The ionic calcium present in aqueous medium is precipitated as calcium oxalate by treating with ammonium oxalate. The amount of calcium is determined titrimetrically by using potassium permanganate solution, in acidic medium.

Reagents:(i) Ammonium oxalate solution 4% (w/v) (ii) Ammonium hydroxide solution (2% v/v) (iii) Potassium permanganate (0.01N) (iv) Sulphuric acid (1N) (v) Concentrated sulphuric acid (vi) Oxalic acid (0.01N) solution.

(**Note**: For preparation of the reagents use deionized or double glass distilled water).

Procedure

(i)   Pipette out 25 mL of oxalic acid solution into a 250 mL conical flask. Add 1 mL of conc. sulphuric acid using a pro-pipette and heat the contents to 70°C on a thermostated hot water bath/ hotplate or a Bunsen burner. Titrate immediately against potassium permanganate solution, taken in a burette, while the content of the conical flask is hot. During initial titration, add potassium permanganate solution drop wise with constant stirring, followed by rapid addition to a pale pink end point. Repeat the titration thrice and use the average value for calculating the normality of potassium permanganate.

(ii)   Take 2 mL of the sample solution into a graduated glass centrifuge tube (15 mL capacity). Add, 2 mL of deionized water followed by 2 mL of ammonium oxalate solution and vortex gently. Leave the centrifuge tube for 30 min at room temperature. Centrifuge the contents at 2500 rpm for 10 min in a table top clinical centrifuge. Discard the supernatant without disturbing the precipitate. Drain the residual supernatant by inverting the tube onto a blotting paper. Wash the precipitate, thrice with 3 mL of 2% ammonium hydroxide solution and vortex gently. After each wash, centrifuge the contents as detailed above. The washing step is essential in order to ensure complete removal of ammonium oxalate.

(iii)   To the washed precipitate, add 2 mL of 1 N H_2SO_4 and mix. Place the tube in a boiling water bath, until all the precipitate is dissolved and then titrate the hot solution with standardized potassium permanganate solution taken in a micro burette to a pale pink end point that persists at least for a minute. Conduct a blank titration using water (2 mL) and 1 N H_2SO_4 (2 mL) only.

Calculation

Sample titer value (mL) - Blank titer value (mL) = X
One mL of 0.01 N potassium permanganate = 0.2004 mg of calcium.

$$\text{Calcium (mg\%)} = \frac{X \times 0.2004}{2} \times 100$$

Workout: Determine the Ca^{++} content in (i) Human/bovine serum (ii) Buffalo milk.

[§] (i) Clark, E. P. & Collip, J. B. J. Biol. Chem. 63: 461, (1925).
   (ii) Sendroy, J. Jr. J. Biol. Chem. 152: 539, (1944).

Experiment #3: Estimation of glucose by *Benedict's* method [§]

Principle: The chemical reaction in olves reduction of *Benedict's* reagent by reducing sugar to a white precipitate of cuprous thiocyanate, in alkaline medium. The disappearance of blue colour is indicative of the end point of reaction.

Reagents: (i) *Benedict's* reagent: This reagent is prepared by dissolving, 200 g of sodium citrate, 75 g of sodium carbonate (anhydrous), 125 g of potassium thiocyanate in 500-600 mL of distilled water, taken in a 1 L glass beaker. To enhance the solubility of the salts, heat the solution on a hot plate. After cooling the salt solution, add 100 mL of cupric sulphate solution (18% w/v) with constant stirring. Later, add 5 mL of (5% w/v) solution of potassium ferrocyanide and make up the volume to 1 L with distilled water, in a volumetric flask (ii) Anhydrous sodium carbonate (solid)

Standard glucose solution (5 mg/mL): Weigh and transfer 500 mg of glucose (anhydrous) into a 100 mL volumetric flask. Dissolve and make up the volume with distilled water.

Procedure

(i) Take 25 mL of *Benedict's* reagent in a 250 mL conical flask and add 3 g of anhydrous sodium carbonate and mix well. Add few pieces of pumice stones or porcelain chips and bring the contents to boiling over a Bunsen flame.

(ii) Titrate rapidly, the contents against the glucose solution taken in a burette. The titration is continued until the blue colour disappears and a white precipitate is obtained. Record the titer value.

Note: Throughout the titration, *Benedict's* reagent should be kept boiling.

Calculation

Twenty five millilitre of *Benedict's* reagent is reduced by 50 mg of glucose. The amount of sugar required to reduce 25 mL of *Benedict's* reagent by other reducing sugars, such as fructose, lactose and maltose are 53, 68.8 and 79 mg, respectively.

Workout: Determine the reducing sugar content in grape juice.

[§] Benedict, S. R. J. Amer. Med. Asso. 57: 1194, (1911).

Experiment #4: Quantitation of glycine by formol titration [§]

Principle: The carboxyl group of the amino acid donates its proton to the amino group to form a zwitterion in solution. The zwitterionic form of the amino acid cannot be accurately titrated to its end point with alkaline indicators. However, *Sørensen* observed that, if the amino acid solution is neutralised and then treated with excess of formaldehyde (neutralised before use), results in a shift of its pK_a value to more acidic region which can be sharply titrated to its end point with alkali. The amount of alkali required corresponds to complete titration of the amino acid, hence it's content.

As the reaction proceeds with formaldehyde, the H^+ is removed from -N^+H_3 by alkali to form NH_2. This is followed by the formation of monomethylol and dimethylol derivatives of the amino acid, in the presence of excess of formaldehyde.

Reagents: (i) Formaldehyde solution: (ii) Standard glycine solution (0.1N) (iii) Phenolphthalein indicator (1% w/v in ethanol) (iv)Standardised sodium hydroxide solution (0.1N).

Procedure

(i)   Take 5 mL of glycine solution in a 100 mL conical flask and add 10 mL of formaldehyde solution along with 2 to 3 drops of indicator and titrate the contents against standardized sodium hydroxide solution, taken in a burette to a pale pink end point.

(ii) Conduct a blank titration using 5 mL of distilled water and 10 mL of formaldehyde solution.

Calculation

(a)   Titer value (mL) of amino acid solution - titer value (mL) of the blank
      $= X$ mL

(b)   1 mL of sodium hydroxide (0.1 N) = 4* mg of NaOH.

$$\text{Concentration of glycine (mg)} = \frac{X \text{ mL} \times 4}{40 \text{ (MW of NaOH)}} \times 75 \text{ (MW of glycine)}$$

---

* The value depends upon the actual normality of the standardized NaOH solution.

Workout: Determine the normality and concentration of test glycine solution.

Note: Normality of glycine as determined by the formol titration multiplied by the molecular weight gives the concentration of glycine in g/L.

[§] (i)   Sørensen, S. P. L. Biochem. Z. 7: 45: (1907).
(ii) Greenstein, J. P. & Winitz, M. *Chemistry of the Amino acids.* Vol. 1. John Wiley & Sons, Inc. NY, USA, (1961).

Experiment #5: Determination of *pK* and *pI* value of an amino acid [§]

Principle: Amino acid at acidic pH exists as cations, anions at alkaline pH and zwitterions at neutral pH. Thus amino acids are amphoteric molecules and can be titrated both by acid and base. In acidic solutions, they are protonated and in basic solutions they are de-protonated. Hence, it is possible to determine the dissociation constants' pK_a (acid), pk_b (base) of the ionizable groups and isoelectric point (pI) of amino acids. The pk_a is a point where the pH of an amino acid in an acidic solution is half protonated, while pk_b is a point where the pH of an amino acid in a basic solution is half deprotonated. An isoelectric point (pI) is the arithmetic mean of pK_a and pK_b value.

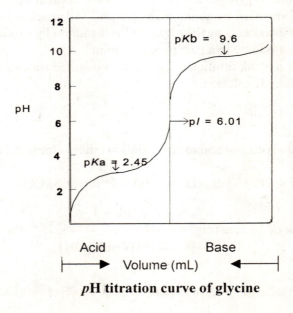

***p*H titration curve of glycine**

Reagents: (i) Standardized sodium hydroxide (0.1 N) (ii) Standardized hydrochloric acid (0.1N) (iii) Phenolphthalein indicator (1% w/v in ethanol).

Amino acid solution: Glycine solution (0.1N).

Equipment: pH meter.

Procedure

(i)   After standardising the acid and base, calculate the volume of alkali and acid required to titrate the amino acid solution by,

$$\frac{\text{Volume of glycine taken for titration (20 mL) x Molarity of glycine (0.1)}}{\text{Molarity of the standardized acid}}$$

$$\frac{\text{Volume of glycine taken for titration (20 mL) x Molarity of glycine (0.1)}}{\text{Molarity of the standardized base}}$$

Half the volume of acid or base consumed during titration gives the pK_a and pK_b values.

(ii) Titrate 20 mL of 0.1 N, (pH adjusted to 6.0), glycine solution taken in a 100 mL conical flask with calculated volume of standardised acid, using a pH meter. Record the change in pH value for every 1 mL addition of acid. Repeat the procedure of amino acid titration with calculated volume of standardised base and record the change in pH value for every 1 mL addition of base.

(iii) Plot on the x- axis the volume of acid and base added *vs* the change in the values of pH on the y- axis. Extrapolate from the x -axis, the value of half the volume of acid or base consumed in the titration to give pK_a and pK_b value. Calculate the isoelectric point of glycine by using the equation,

$$(pI) = \tfrac{1}{2}(pK_a + pK_b)$$

Workout: Determine the *pI* value of alanine.

[8] (i)   Greenstein, J. P. J. Biol. Chem. 93: 279, (1931).
   (ii) Greenstein, J. P. & Winitz, M. *Chemistry of the Amino acids*. Vol. 1. John Wiley & Sons, Inc. NY, USA, (1961).

Experiment #6: Determination of ion-exchange capacity of a resin.

Principle: The exchange capacity determination involves displacement of H^+ ions of the charged groups of an ion-exchange matrix with sodium ions followed by titration of the acid with standardised alkali using phenolphthalein indicator.

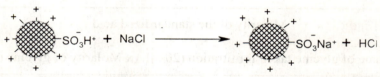

Ion-exchange resin - Hydrogen ion form   Ion-exchange resin Sodium form

Reagents: (i) Sodium hydroxide (4 N) (ii) Hydrochloric acid (4 N) (iii) Sodium chloride(1 M) (iv) Standardized 0.1 N sodium hydroxide (v) Phenolphthalein indicator.

Ion-exchange resin (cation exchanger): Dowex-50 H^+ (200-400 mesh). Chemically, Dowex resin is sulphonated polystyrene cross-linked with divinyl benzene.

Procedure

(i)   Dowex-50 is commercially available either as H^+ or Na^+ form. The later form cannot be directly employed for exchange capacity determination. It has to be converted to the H^+ form.

(ii) Prior to experimentation, the resin has to be thoroughly washed. Take 20 g of the resin in a 250 mL glass beaker and wash successively with 4 N NaOH (100mL, once), distilled water (100 mL, thrice), 4 N HCl (100mL, once), followed by several washes of distilled water to remove acid. Later the resin is filtered through a sintered funnel, followed by methanol (100-150 mL) wash. The washed resin is dried overnight at room temperature.

(iii) Weigh 1g of the dried resin (H^+ form) into a 100 mL beaker. Add 20 mL of 1M NaCl solution to the flask and leave the contents for 45 min with intermittent shaking so as to facilitate the ion-exchange process.

(iv) Filter the contents into a 150 mL conical flask using a glass funnel plugged with glass wool. Rinse the beaker, twice with 5 mL distilled water and transfer the rinses to the funnel, so as to ensure the quantitative transfer. Further, wash the resin with 5 mL of distilled water and collect the rinse into the conical flask.

(v) To the filtrate in the conical flask, add 2-3 drops of phenolphthalein indicator and titrate the contents against standardized 0.1 N NaOH, taken in a glass burette, until a pale pink colour appears. Note the volume of NaOH consumed (titer value in mL). Perform a blank titration with 20 mL of 1 M sodium chloride solution and note the titer value.

Calculations

(i) The exchange capacity of resin is expressed in terms of milliequivalents (mEq) per gram of resin.

(ii) One mL of 0.1 NaOH is equivalent to 4 mg by weight.

(iii) One milliequivalent (mEq) of NaOH = 40 mg

(iv) Actual titer value = Titer value of resin (mL) - Blank titer value (mL)
= X mL

$$\text{Exchange Capacity (mEq g}^{-1}) = \frac{X \text{ mL} \times 4}{40 \text{ (1 mEq of NaOH)}}$$

Workout: Compare the exchange capacity of resin (H^+) obtained from two different manufacturers.

Experiment # 7: Determination of acid value of fats [§]

Principle: The acid value is defined as milligrams of KOH required to neutralise the free fatty acid present in one gram of fat. The free fatty acids are generated upon prolonged storage due to oxidation and hydrolysis of fats. Acid value is an index of keeping quality of fat.

Reagents: (i) Fat solvent: Mix, ethanol (95% v/v) and diethyl ether in the ratio of 1:1 (ii) Potassium hydroxide solution (0.1N) (iii) Phenolphthalein indicator (1% w/v in ethanol).

Procedure

(i)   Standardisation of alkali: Potassium hydroxide (0.1 N) being a secondary standard, determine its normality by titrating with standard potassium hydrogen phthalate (0.1N) or oxalic acid (0.1N). Take 5 mL of potassium hydrogen phthalate (0.1 N) along with two drops of phenolphthalein indicator in a 50 mL conical flask and titrate against potassium hydroxide solution (0.1 N) using a glass burette. A pale pink colour indicates the end point. Calculate the actual normality of the KOH solution prepared.

$$\text{Normality of KOH} = \frac{\text{Volume of phthalate (mL) x normality of phthalate}}{\text{Volume of KOH consumed (mL)}}$$

(ii)   Weigh accurately 10 g of fat into a clean and dry 250 mL conical flask. Add 100 mL of the fat solvent along with 2 - 3 drops phenolphthalein indicator and mix. Titrate against the standardised KOH, to a pale pink end point. Note the titer value and calculate the acid value of the fat.

Calculation:

$$\text{Acid value} = \frac{\text{Titer value x Normality of Standardised KOH x MW of KOH (56.1)}}{\text{Weight of fat (g)}}$$

Workout: Determine the acid value of the following: (i) Coconut oil (ii) Butter (iii) Safflower oil. Compare the acid values of these fats.

§ Official & Tentative Methods of the American Oil Chemist's Society. AOCS, 3rd ed. Illinois, USA, (1981).

Experiment #8: Determination of saponification value of fats §

Principle: Fats (triglycerides) upon alkaline hydrolysis (either with KOH or NaOH) yield glycerol and K^+ or Na^+ - salts of fatty acids (soap). Thus saponification number is milligrams of KOH required to saponify the fatty acids librated from 1 g of fat.

| Triglyceride | Alkali | Glycerol | Potassium salt of fatty acid |

Reagents: (i) Ethanolic KOH (95 % ethanol, v/v; potassium hydroxide, 0.5 N) (ii) Phenolphthalein indicator (1 % w/v in ethanol) (iii) Hydrochloric acid 0.5 N.

Note: Standardize the acid strength (HCl) using standardised KOH solution, as detailed in Experiment 7.

Procedure

(i)　Weigh 1g of fat in a 50 mL glass beaker. Add 5 mL of ethanolic KOH to the beaker so as to dissolve the fat. Transfer the contents of the beaker into a 250 mL Erlenmeyer flask (conical flask). Rinse the beaker twice, with 5mL of ethanolic KOH to ensure quantitative transfer of the fat. Further, add 10 mL of ethanolic KOH to the conical flask and reflux the contents with a reflux condenser (water or air condenser) attached through a one holed bark cork, over a boiling water bath for 1 hour. Simultaneously, setup a blank, containing 25 mL of ethanolic KOH which is devoid of fat sample, and reflux for 1 hour.

(ii)   Cool the contents of fat sample and blank to room temperature and titrate against 0.5 N HCl, using one or two drops of phenolphthalein indicator. Record the end point and calculate the saponification value of the oil sample as follows,

[Volume of 0.5 N HCl consumed (in mL) for blank - Volume of 0.5 N HCl consumed (mL) for fat sample] = volume of KOH (mL) consumed by 1 g fat. Each mL of 0.5 N alkali is equivalent to 28.05 mg of KOH.

Saponification value or number of fat = Millilitres of 0.5 N KOH consumed by 1 g fat X 28.05

Workout: Determine the saponification value of (i) Coconut oil (ii) Peanut oil and (iii) Butter fat. Compare the values.

[§] Official & Tentative Methods of the American Oil Chemist's Society. AOCS, 3rd ed. Illinois, USA, (1981).

Experiment #9: **Determination of iodine number of oil** [§]

Principle: Halogens such as iodine, bromine add across the double bonds of unsaturated oils to form saturated halo-products. The extent of halogenation is a measure of the degree of unsaturation in oils. Thus, iodine number is defined as grams of iodine absorbed by 100 g of fat.

Unsaturated fatty acid

Di-halo adduct of fatty acid

The reaction is conducted with iodine monochloride which reacts with the unsaturated fat. The unreacted iodine monochloride is determined after converting it to free iodine (I_2) in the presence of potassium iodide. The librated iodine is titrated against sodium thiosulphate.

(i) .... $ICl + KI \rightarrow KCl + I_2$
(Iodine monochloride)

(ii) .... $2Na_2S_2O_3 + I_2 \rightarrow 2NaI + Na_2S_4O_6$
(Sodium thiosulphate)   (Sodium dithionite)

Reagents: (i) Iodine monochloride reagent: Transfer 5 mL of iodine monochloride into a 500 mL volumetric flask containing 100 mL of glacial acetic acid. Make up the volume to 500 mL with acetic acid (ii) Potassium iodide (10 % w/v; prepare this reagent freshly) (iii) Sodium thiosulphate 0.1 N (freshly prepared) (iv) Starch indicator solution (1% w/v): Weigh 1 g of soluble starch and transfer it into a 50 mL beaker. Prepare a paste by adding distilled water. Boil 80 mL of distilled water in a glass beaker and transfer the starch paste in small proportions using a policeman glass rod. After cooling to room temperature, make up the volume to 100 mL. Transfer and store it in a reagent bottle layered with toluene (v) Fat sample (2% w/v) in chloroform.

(**Caution**: Commercially, iodine monochloride is available in sealed glass vials. Handle the chemical with disposable gloves (corrosive chemical). Pre-cool the glass vial before using. Cut the tip of the vial using a glass cutting file in a fume hood and transfer the required volume and store the unused iodine monochloride, safely).

Procedure

(i)   Pipette out 10 mL of the fat sample into an iodination flask (250 mL). Add 20 mL of iodine monochloride reagent. Stopper the flask and shake the contents. Leave the flask in dark at room temperature for 30 min. Separately, setup a reagent blank without fat sample.
Caution: No mouth pipetting of iodine monochloride reagent.

(ii) Add 10 mL of potassium iodide solution to the flask and mix the contents. Rinse the stopper using 30 - 50 mL distilled water. The contents of the flask are titrated against sodium thiosulphate (0.1 N) solution, taken in a glass burette, until a pale straw colour is observed. Add 1 - 2 mL of starch indicator. Addition of the indicator results in blue colouration. Continue the titration until the solution turns colourless. Note the titer values, both for, fat sample and blank. The difference between the blank and the fat sample gives the amount of iodine absorbed by the fat.

Calculation

Each mL of sodium thiosulphate (0.1 N) is equivalent to 12.7 mg of iodine.

$$\text{Iodine number of fat} = \frac{X\,(\text{mL}) \times 12.7}{\text{Weight of fat sample used for analysis (g)}} \times 100$$

Where, X is difference between blank and sample titer value.

Workout: Determine the iodine number of (i) Coconut oil (ii) Sun flower oil (iii) Hydrogenated oil (Dalda / Vanaspathi). Rank the degree of unsaturation in the sample analysed.

§ Official & Tentative Methods of the American Oil Chemist's Society. AOCS, 3rd ed. Illinois, USA, (1981).

*(see, **appendix** for Acid, Saponification & Iodine values of common oils & fats).*

Experiment #10: Determination of peroxide value of oil §

Principle: Peroxide value is an indication of the extent of oxidation suffered by an oil. The peroxides present in such oils are determined by its titration against sodium thiosulphate in the presence of potassium iodide and starch, which is used as an indicator.

Reagents: (i) Solvent mixture: Mix three volumes of glacial acetic acid and one volume of chloroform (ii) Freshly prepared saturated potassium iodide solution (iii) Sodium thiosulphate solution 0.01 N (earlier standardized against potassium dichromate solution) (iv) Starch indicator: Soluble starch (1% w/v) solution.

Procedure

(i)   Accurately weigh 5 g of test oil sample into a 250 mL stoppered conical flask. Add 30 mL of solvent mixture and mix. Add 0.5 mL saturated potassium iodide solution and mix. Allow the contents of conical flask to stand in dark for 1 min with occasional shaking, followed by addition of 30 mL of distilled water. Titrate the librated iodine with sodium

B. SPECTROPHOTOMETRIC & PHOTOMETRIC EXPERIMENTS

Amino acid analysis

Experiment #12: Determination of molar absorption coefficient (ε) of L-tyrosine.

Principle: Molar absorption coefficient (ε) is defined as the absorbance of a one molar solution of pure absorbing material in a 10 mm path length cell, under specified conditions of wavelength (λ_{max}) and solvent. The units of ε are $M^{-1} cm^{-1}$. Molar absorption coefficient value of a biomolecule can be determined by using the equation $A = \varepsilon\, cl$. Experimentally, absorbance value of a known concentration of the pure analyte is obtained by recording the absorbance at its λ_{max}, using a UV-Vis spectrophotometer.

Equipment: UV-Vis spectrophotometer, analytical balance.

Reagent: (i) Prepare 5×10^{-4} M (or mole/L) solution of L-tyrosine in 0.1 N NaOH (ii) Sodium hydroxide solution (0.1 N) (prepared by using fresh double glass distilled water).

Procedure

(i)   Turn on the UV-Vis spectrophotometer and allow it to warm up and auto-calibrate.

(ii) Select the wavelength to record the absorbance ($\lambda_{max} = 293.5$ nm) of the prepared tyrosine solution.

(iii) Set the instrument in the absorbance mode and calibrate the instrument to zero by placing a blank (0.1 NaOH solution) taken in a 3 mL (1 cm path length) quartz cuvette.

(iv) Record the absorbance of the prepared tyrosine solution at 293.5 nm, taken in the quartz cuvette.

Calculation

Substitute the experimental value of A. recorded from the spectrophotometer in the equation: $A = \varepsilon\ cl$, and calculate the molar absorption coefficient of L-tyrosine.

$$\text{Molar absorption coefficient } (\varepsilon) = A \div l \times c = \frac{A \text{ (recorded at } \lambda_{293.5\,nm})}{(1 \text{ cm}) (5 \times 10^{-4} \text{ mole/L})}$$

Workout: What is the molar absorption coefficient value of 5×10^{-4} mole/L solution (prepared in 0.1 N NaOH) of L- tryptophan? (Take tryptophan λ_{max} (in 0.1 N NaOH) as 280 nm and path length = 1 cm).

Experiment #13: Estimation of amino acid by ninhydrin method [§]

Principle: Ninhydrin, also chemically known as triketohydrindene hydrate reacts with amino acids to give a purple coloured complex (*Ruhemann's purple*) with an absorption maximum at 570 nm. However, imino acids such as proline and hydroxy-proline yield a yellow colour with an absorption maximum at 440 nm.

Ninhydrin oxidises the amino acid to aldehyde, releasing carbondioxide and ammonia. During the course of the reaction, ninhydrin gets reduced to hydridantin. The hydridantin formed condenses with ninhydrin in the presence of ammonia to yield a purple coloured complex - *Ruhemann's purple*.

Note: Primary amines also react with ninhydrin, however there in no liberation of CO_2.

| Amino acid | Ninhydrin | | Ruhemann's purple | Aldehyde |

Standard amino acid stock solution (10 μ moles/mL or 1.31 mg/mL): (i) Weigh 131 mg of leucine and transfer it carefully into a 100 mL volumetric flask. Suspend the amino acid in 20-30 mL of distilled water and warm the contents in a hot water bath for 10 min. After cooling, make up the volume to 100 mL with distilled water (ii) Working standard solution (1 μ mole/mL or 131μg/mL): Dilute 10 mL of the stock solution to 100 mL with distilled water in a volumetric flask.

Reagents: (i) Citrate buffer 0.2 M, *p*H 5.0: Dissolve 10.51 g of citric acid (monohydrate) and 14.71 g of sodium citrate(dihydrate) in 150 mL of distilled water. Adjust the *p*H of the solution to 5.0, by adding sodium hydroxide solution (1N) and make up the volume to 250 mL with distilled water, in a volumetric flask. (ii) Ninhydrin reagent: Dissolve 0.4 g of stannous chloride (monohydrate) in 250 mL of 0.2 M citrate buffer, *p*H 5.0. To this solution add 50 mL of methyl cellosolve (ethylene glycol monomethyl ether), containing 1 g of ninhydrin and mix the contents. Store the reagent in a brown bottle. Use freshly prepared ninhydrin reagent for analysis. (iii) Diluent solution: Mix *n*-propyl alcohol and distilled water in equal proportions (1:1).

Note: (i) Ninhydrin reagent, contains in addition to ninhydrin, stannous chloride which acts as a reducing agent. Use peroxide free methyl cellosolve. This can be tested by mixing 3 mL of cellosolve with an equal volume of 4% potassium iodide solution. Development of yellow colour indicates the presence of peroxides. (ii) Ninhydrin is a toxic compound. Wear disposable gloves while handling the chemical.

Procedure

(i)   To 1.0 mL of the standard solution (containing 0.1 - 1.0 μ mole of L-leucine) or appropriately diluted unknown sample solution, add 2.0 mL of ninhydrin reagent. Heat the test tubes covered with glass marbles, in a boiling water bath for 15 min. Cool the tubes to room temperature and add 7 mL of diluent solution and vortex.

(ii)   Measure the purple colour developed against the reagent blank at 570 nm in a photometer/colorimeter and record the absorbance.

(iii) Construct a calibration curve on a graph paper, by plotting the leucine concentration (0.1 - 1.0 μmole) on *x*-axis and absorbance at 570 nm on the *y*-axis. Compute the concentration of the amino acid in the sample from the calibration curve.

Quantitative Analysis

Protocol:

S.No	Amino acid Standard (mL)	Distilled water (mL)	Conc. of Amino acid (µ mole)	Ninhydrin reagent (mL)	Diluent solution (mL)	Absorbance at 570 nm
1	Blank	1.0	- -	2. 0	7. 0	
2	0.1	0.9	0.1	"	"	
3	0.2	0.8	0.2	"	"	
4	0.3	0.7	0.3	"	"	
5	0.4	0.6	0.4	"	"	
6	0.5	0.5	0.5	"	"	
7	0.6	0.4	0.6	"	"	
8	0.7	0.3	0.7	"	"	
9	0.8	0.2	0.8	"	"	
10	0.9	0.1	0.9	"	"	
11	1.0	0.0	1.0	"	"	
12	Sample-1		(to be determined)	"	"	
13	Sample-2		(to be determined)	"	"	

" = same volume

Workout: (a) Construct the calibration curves for (i) lysine (0.1 - 1.0 µ mole) (ii) glutamic acid (0.1 - 1.0 µ mole) and compare the colour yield with standard leucine.

§ (i) Ruhemann. S. J. Chem. Soc. 99: 1491, (1911).
  (ii) Moore, S. & Stein, W. H. J. Biol. Chem. 176: 367, (1948).
  (iii) Mac Fadyn, D. A. J. Biol. Chem. 186: 1, (1950).
  (iv) Rosen, H. Arch. Biochem. Biophy. 67: 10, (1957).
  (v) Colowick, S. P. & Kaplan, N. O. (eds). *Methods in Enzymology*. Vol. VI. Academic Press, NY, USA, (1963).

Experiment #14: Determination of tyrosine by nitrosonaphthol method[§]

Principle: The phenolic group of tyrosine reacts with nitrosonaphthol under acidic conditions to yield a red coloured complex, which has an absorbance maximum at 520 nm. The colour developed is stable and the method is sensitive, reliable and can be directly used for the estimation of tyrosine content in proteins.

Standard tyrosine stock solution (1.5mg/mL): (i) Weigh 150 mg of tyrosine and transfer it carefully into a 100 mL volumetric flask. Suspend the amino acid in 20-30 mL of distilled water. Dissolve the amino acid by adding 2 mL of 5 N HCl and make up the volume to 100 mL with distilled water (ii) Working standard tyrosine solution (150 μg/mL): Dilute 10 mL of the stock solution to 100 mL with distilled water in a volumetric flask.

Reagents: (i)Nitrosonaphthol reagent : Nitrosonaphthol (0.15% w/v) in 0.1 N sodium hydroxide solution. Filter before use. (ii) Acid-base reagent : Mix, equal volumes of nitric acid (0.025 N) and sodium hydroxide (0.3 N) (this reagent to be prepared freshly) (iii) Concentrated sulphuric acid (Analar grade).

Procedure

(i)   To one millilitre of the standard solution (containing 10 - 150 μg of tyrosine) or appropriately diluted unknown sample solution, add 1 mL of nitrosonaphthol reagent and 2 mL of acid-base reagent and mix. Heat the tubes in a boiling water bath for 10 minutes. Cool the tubes to room temperature and add 4 mL of concentrated sulphuric acid drop wise, using a burette or a pro-pipette and vortex.

(ii) Measure the red colour developed against the reagent blank at 520 nm in a photometer/colorimeter and record the absorbance.

(iii) Construct a calibration curve on a graph paper, by plotting the tyrosine concentration (10-150 μg) on *x*- axis and absorbance at 520 nm on the *y*- axis. Compute the concentration of the tyrosine in the sample from the calibration curve.

Protocol:

S.No	Standard tyrosine (mL)	Distilled water (mL)	Conc. of tyrosine (μg)	Nitrosona-pthol reagent (mL)	Acid-base reagent (mL)	Conc. H$_2$SO$_4$ (mL)	Absorbance at 520nm
1	Blank	1.0	- -	1.0	2.0	4.0	
2	0.1	0.9	15	"	"	"	
3	0.2	0.8	30	"	"	"	
4	0.3	0.7	45	"	"	"	
5	0.4	0.6	60	"	"	"	
6	0.5	0.5	75	"	"	"	
7	0.6	0.4	90	"	"	"	
8	0.7	0.3	105	"	"	"	
9	0.8	0.2	120	"	"	"	
10	0.9	0.1	135	"	"	"	
11	1.0	- -	150	"	"	"	
12	Sample - 1		(to be deter-mined)	"	"	"	
13	Sample - 2		(to be deter-mined)	"	"	"	

" Same volume

Workout: Determine the tyrosine content of (i) Casein (ii) Bovine serum albumin.

Note: Prepare 5 - 10 mg/mL of protein solution for analysis.

[§] Uehara, K., Mannen, S. & Kishida, K. J. Eiochem. 68: 119, (1970).

Experiment #15: Estimation of tryptophan by *Spies & Chamber's* method [s]

Principle: Tryptophan condenses with p - dimethylaminobenzaldehyde (*Ehrlich's* reagent) under acidic conditions to a colourless condensation product, which upon further oxidation with sodium nitrite yields a blue colour that has an absorption maximum at 590 nm.

Standard tryptophan stock solution (1mg/mL): (i) Weigh 100 mg of tryptophan and transfer it carefully into a 100 ml volumetric flask. Suspend the amino acid in 20-30 mL of distilled water. Dissolve the amino acid by adding 2 mL of 5N HCl and make up the volume to 100mL with distilled water (ii) Working standard tryptophan solution (100 µg/mL): Dilute 10 mL of the stock solution to 100 mL with distilled water in a volumetric flask.

Reagents: (i) *Ehrlich's* reagent : p - dimethylaminobenzaldehyde (3% w/v) in 2 N HCl. Filter before use. (ii) Sulphuric acid 23.7 N (**Note**: Prepare the solution in a glass beaker, placed in ice cold water by adding acid to distilled water in small volumes) (ii) Sodium nitrite ($NaNO_2$) solution: Sodium nitrite (0.045 %w/v) in distilled water (to be freshly prepared).

Procedure

(i)   To one millilitre of the standard solution (containing 10 - 100 µg of tryptophan) or appropriately diluted or undiluted unknown sample solution, add 1 mL of *Ehrlich's* reagent and 8 mL of sulphuric acid (23.7 N), using a burette and mix. Incubate the tubes at room temperature, in dark for 60 min. After incubation, add 0.1 mL of sodium nitrite solution and vortex. Allow the tubes to stand for 30 min at room temperature for colour development.

(ii)   Measure the blue colour against the reagent blank at 590nm in a photometer/colorimeter and record the absorbance.

(iii) Construct a calibration curve on a graph paper, by plotting the tryptophan concentration (10 - 100 µg) on x- axis and absorbance at 590 nm on the y- axis. Compute the concentration of the tryptophan in the sample from the calibration curve.

Protocol:

S.No	Standard tryptophan (mL)	Distilled water (mL)	Conc. of tryptophan (μg)	Ehrlich's reagent (mL)	Sulphuric acid (23.7N) (mL)	NaNO$_2$ solution (mL)	Absorbance at 590nm
1	Blank	1.0	- -	1.0	8.0	0.1	
2	0.1	0.9	10	"	"	"	
3	0.2	0.8	20	"	"	"	
4	0.3	0.7	30	"	"	"	
5	0.4	0.6	40	"	"	"	
6	0.5	0.5	50	"	"	"	
7	0.6	0.4	60	"	"	"	
8	0.7	0.3	70	"	"	"	
9	0.8	0.2	80	"	"	"	
10	0.9	0.1	90	"	"	"	
11	1.0	- -	100	"	"	"	
12	Sample - 1		(to be determined)	"	"	"	
13	Sample - 2		(to be determined)	"	"	"	

" Same volume

Workout: Estimate the tryptophan content of Bovine serum albumin and determine the tyrosine-tryptophan ratio in the protein.

§(i) Spies, J. R., & Chambers, D.C. Anal. Chem. 20: 30, (1948)
(ii) Spies, J. R., & Chambers, D.C. Anal. Chem. 21: 1249, (1949)

Protein analysis

Experiment #16: Determination of protein by biuret method [s]

Principle: The peptide bonds of the protein react with cupric copper (Cu^{2+}) under alkaline condition to yield a purple / violet coloured complex, which shows an absorption maximum at 540 nm.

Purple coloured peptide-copper coordination complex

Biuret Reaction

Protein reference standard: Casein is used as a reference standard protein (10 mg/mL). Weigh accurately 1 g of casein and transfer it to a clean and dry 100 mL volumetric flask. Suspend the protein in 50 mL of distilled water. Dissolve the protein by adding few drops of 10 N sodium hydroxide solution and make up the volume to 100 mL.

Reagents: Biuret reagent: Dissolve 1.5 g of cupric sulphate in 250 mL distilled water. Separately, weigh and dissolve 6 g of sodium potassium tartrate in 250 mL of distilled water. Mix the cupric sulphate and tartrate solutions in a 1 L beaker. To this solution add 300 mL of 10% (w/v) sodium hydroxide solution, with constant stirring. Make up the volume to 1L with distilled water using a volumetric flask. Store the biuret reagent in a plastic container.

Procedure

(i) To one millilitre of the standard protein solution (containing 1 - 10 mg of casein) or appropriately diluted or undiluted unknown protein sample solution, add 5 mL of biuret reagent and mix the contents.

(ii) After 30 min of incubation at room temperature, measure the violet colour developed against the reagent blank at 540 nm in a photometer/ colorimeter and record the absorbance.

(iii) Construct a calibration curve on a graph paper, by plotting the protein concentration (1-10 mg of protein) on x- axis and absorbance at 540 nm on the y- axis. Compute the concentration of the protein in the sample from the calibration curve. While calculating the protein concentration in the unknown sample, the dilution factor has to be taken into account.

Protocol:

S.No	Standard protein (mL)	Distilled water (mL)	Conc. of casein (mg)	Biuret reagent (mL)	Absorbance at 540 nm
1	Blank	1.0	- -	5.0	
2	0.1	0.9	1	"	
3	0.2	0.8	2	"	
4	0.3	0.7	3	"	
5	0.4	0.6	4	"	

6	0.5	0.5	5	,,	
7	0.6	0.4	6	,,	
8	0.7	0.3	7	,,	
9	0.8	0.2	8	,,	
10	0.9	0.1	9	,,	
11	1.0	- -	10	,,	
12	Sample-1		(to be determined)	,,	
13	Sample-2		(to be determined)	,,	

" = same volume

Note: Ammonium salts, Tris, sucrose and glycerol interfere and affect the development of colour given by proteins. Presence of lipids and detergents may cause turbidity.

Workout: Estimate the protein content in egg-white (hen's egg) and report the value of protein as g%.

§ Gornall. A.G. Bardawill, C. S. & David, M. M. J. Biol. Chem. 177: 751, (1949).

Experiment #17: Estimation of protein by *Lowry* method §

Principle: The blue colour developed in Lowry method is a result of (i) reaction of the peptide bonds of the protein with cupric copper under alkaline conditions and (ii) reduction of phosphomolybdic acid by tyrosine and tryptophan (aromatic amino acids) residues of the protein. The colour shows a λ_{max} at 720 nm.

Protein reference standard: Bovine serum albumin (BSA) is used as a reference standard protein. Weigh accurately 100 mg of BSA and transfer it into a clean and dry 100 mL volumetric flask. Dissolve the protein by adding 5 - 10 mL of 0.1N sodium hydroxide and make up the volume to 100 mL with distilled water. Dilute the protein solution (1 mg/mL) to give a working standard concentration of 100 μg/mL (10 X dilution) in a volumetric flask.

Reagents

(i)   Sodium carbonate (2 % w/v) in 0.1N Sodium hydroxide solution (**Reagent A**) : To prepare 1 L of the reagent, dissolve 20 g of sodium carbonate and 4 g of sodium hydroxide in distilled water and make up the volume to 1 L, using a volumetric flask.

(ii) Copper sulphate solution (1 % w/v) (**Reagent B**) : Dissolve 1 g of copper sulphate in distilled water and make up the volume to 100 mL.

(iii) Sodium potassium tartrate solution (2 % w/v) (**Reagent C**): Dissolve 2 g of the salt in distilled water and make up the volume to 100 mL.

(iv) Alkaline copper reagent (**Reagent D**) : To 100 mL of reagent A. add 2 mL of reagent B & C (mix 1 mL, each of B & C in a glass test tube and vortex) and stir the solution and use it for analysis. This reagent should be prepared freshly.

(v)   Folin's reagent (1 N strength): Check the normality of the commercial Folin's reagent and dilute appropriately, before use.

Procedure

(i)   To one millilitre of the standard protein solution containing 10 - 100 μg of protein or appropriately diluted unknown protein sample solution, add 4 mL of **reagent D** and mix the contents.

(ii) After 10 min of incubation at room temperature, add 0.4 mL of Folin's reagent and vortex the contents immediately.

(iii) Run a reagent blank with 1mL of distilled water along with the standards and sample.

(iv) After 30 min of incubation at room temperature, read the blue colour developed at 720 nm, in a photometer and record the absorbance.

(v)   Construct a calibration curve on a graph paper, by plotting the protein concentration (10 - 100 μg of protein) on *x*-axis and absorbance at

720 nm on the y- axis. Compute the concentration of the protein in the sample from the calibration curve. While calculating the protein concentration in the unknown sample, the dilution factor has to be taken into account.

Protocol:

S.No	BSA Standard (mL)	Distilled water (mL)	Conc. of protein (μg)	Reagent D (mL)	Folin's Reagent (mL)	Absorbance at 720 nm
1	Blank	1.0	- -	4.0	0.4	
2	0.1	0.9	10	"	"	
3	0.2	0.8	20	"	"	
4	0.3	0.7	30	"	"	
5	0.4	0.6	40	"	"	
6	0.5	0.5	50	"	"	
7	0.6	0.4	60	"	"	
8	0.7	0.3	70	"	"	
9	0.8	0.2	80	"	"	
10	0.9	0.1	90	"	"	
11	1.0	- -	100	"	"	
12	Sample-1		(to be determined)	"	"	
13	Sample-2		(to be determined)	"	"	

" = same volume

Note: (i) For a good analytical result, it is preferable to run the reagent blank, standards and the unknown samples in duplicates. (ii) To avoid frothing, do not shake the protein solution vigorously (iii) Ammonium sulphate interferes with the protein assay. It can be removed from the protein solution either by dialysis or gel filtration.

Interfering substances: In Lowry method, the colour yields are affected by a variety of interfering substances, such as free tyrosine and tryptophan, phenolic substances, buffers (Tris, citrate), glucose, sucrose, glycerol, thiol compounds, EDTA, ammonium salts and detergents (Triton-X 100, Chaps).

Workout: Estimate the total protein content in bovine serum or human serum sample.

[§] Lowry, O. H. Rosebrough, N. J. Farr, A. L. & Randall, R. J., J. Biol. Chem. 193: 265, (1951).

Carbohydrate analysis

Experiment #18: Quantitation of total sugars by anthrone method [§]

Principle: Sugars in the presence of sulphuric acid get dehydrated to furfural or hydroxymethyl furfural, which later react with anthrone (dihydro-oxoanthracene) to yield a bluish-green colour complex that has an absorption maximum at 620 nm.

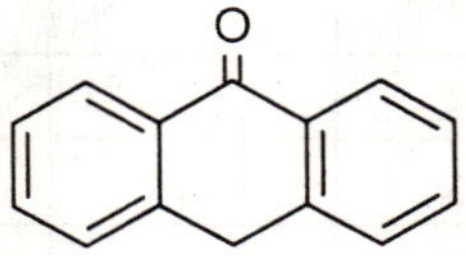

Anthrone

Standard glucose stock solution (1 mg/mL): (i) Weigh 100 mg of glucose and transfer it carefully into a 100 mL volumetric flask. Make up the volume to 100 mL with distilled water (ii) Working standard glucose solution (100 µg/mL): Dilute 10 mL of the stock solution to 100 mL with distilled water in a volumetric flask.

Reagent: Anthrone reagent: Anthrone (0.2 % w/v) in concentrated sulphuric acid.

Procedure:

(i)   To one millilitre of the standard solution (containing 10 - 100 µg of glucose) or appropriately diluted unknown sugar sample solution, add 4 mL of anthrone reagent rapidly into the test tubes using a burette or a pro-pipette and vortex.

(ii) Heat the tubes in a boiling water bath for 10 min. Cool the tubes to room temperature and read the bluish-green colour developed against reagent blank at 620nm in a photometer / colorimeter and record the absorbance.

(iii) Construct a calibration curve on a graph paper, by plotting the glucose concentration (10-100 µg) on *x*- axis and absorbance at 620 nm on the *y*- axis. Compute the concentration of the sugar in the sample from the calibration curve. While calculating the sugar concentration in the unknown sample, the dilution factor has to be taken into account.

Protocol:

S.No	Standard Glucose (mL)	Distilled water (mL)	Conc. of Glucose (µg)	Anthrone reagent (mL)	Absorbance at 620 nm
1	Blank	1.0	- -	4.0	
2	0.1	0.9	10	,,	
3	0.2	0.8	20	,,	
4	0.3	0.7	30	,,	
5	0.4	0.6	40	,,	
6	0.5	0.5	50	,,	
7	0.6	0.4	60	,,	
8	0.7	0.3	70	,,	
9	0.8	0.2	80	,,	
10	0.9	0.1	90	,,	
11	1.0	- -	100	,,	
12	Sample -1		(to be determined)	,,	
13	Sample -2		(to be determined)	,,	

,, = same volume

Quantitative Analysis

Workout: Determine the sugar content of commercial honey sample by anthrone method.

[§] (i) Scott, T. A. & Melvin, E. H. Anal. Chem. 25: 1656, (1953).

Experiment #19: Estimation of total sugars by phenol sulphuric method[§]

Principle: Sugars undergo dehydration in the presence of sulphuric acid to furfural or hydroxymethyl furfural that condense with phenol to form a yellowish-orange coloured compound with an absorption maxima at 490 nm.

Standard stock sugar solution (1 mg/mL): (i) Weigh 100mg of mannose or galactose or glucose and transfer it carefully into a 100 mL volumetric flask. Make up the volume to 100 mL with distilled water (ii) Working standard sugar solution (100 µg/mL): Dilute 10 mL of the stock solution to 100 mL with distilled water in a volumetric flask.

Reagents (i) Phenol reagent: Phenol (5 % w/v) in distilled water (ii) Concentrated sulphuric acid.

Note: Phenol used for the reagent should be colourless. In case it is coloured, the phenol should be distilled.**Caution**: Phenol (a corrosive chemical) distillation should be carried out only under the supervision of the instructor.

Procedure

(i)   To 2 mL of the standard solution (containing 10 -100 µg of sugar) or appropriately diluted unknown sugar sample solution, add 1 mL of phenol reagent, followed by rapid addition of 5 mL of concentrated sulphuric acid, using a burette or a pro-pipette and vortex.

(ii)   After 30 min read the yellowish-orange colour developed against reagent blank at 490 nm in a photometer/colorimeter and record the absorbance.

(iii)   Construct a calibration curve on a graph paper, by plotting the sugar concentration (10 - 100 µg) on x - axis and absorbance at 490 nm on the *y*- axis. Compute the concentration of the sugar in the sample from the calibration curve. While calculating the sugar concentration in the unknown sample, the dilution factor has to be taken into account.

Protocol:

S.No	Standard sugar (mL)	Distilled water (mL)	Conc. of sugar (μg)	Phenol Reagent (mL)	Conc. Sulphuric acid (mL)	Absorbance at 490 nm
1	Blank	2.0	- -	1.0	5.0	
2	0.1	1.9	10	"	"	
3	0.2	1.8	20	"	"	
4	0.3	1.7	30	"	"	
5	0.4	1.6	40	"	"	
6	0.5	1.5	50	"	"	
7	0.6	1.4	60	"	"	
8	0.7	1.3	70	"	"	
9	0.8	1.2	80	"	"	
10	0.9	1.1	90	"	"	
11	1.0	1.0	100	"	"	
12	Sample -1		(to be determined)	"	"	
13	Sample -2		(to be determined)	"	"	

" = same volume.

Workout: Analyse the sugar content of (i) Sugarcane juice (b) Ovalbumin. (This method can also be used for determination of carbohydrate content in glycoproteins).

$ Dubois, M. Gilles, K. A. Hamilton, J. K. Rebers, P. A. & Smith, F. Anal. Chem. 28: 350, (1956).

Experiment #20: Estimation of reducing sugars by dinitrosalicylate (DNS) method [§]

Principle: Reducing sugars convert dinitrosalicylate under alkaline condition to amino-nitrosalicyalate, an orange-yellowish compound that has an absorption maximum at 540 nm.

Dinitrosalicylate   3-amino-5-nitrosalicylate

Standard maltose solution (2 mg/mL): (i) Weigh 200 mg of maltose and transfer it into a volumetric flask, dissolve and make up the volume to 100 mL, with distilled water.

Reagents: (i) DNS reagent: Dissolve 1 g of DNS, 200 mg of phenol and 50 mg of sodium sulphite in 100 mL of 1 % (w/v) NaOH. Store this reagent in a refrigerator (ii) Potassium-sodium tartrate tetrahydrate (40% w/v)(*Rochelle salt*).

Procedure

(i)   To 1 mL of the sugar solution (containing 0.2 - 2 mg of maltose) or appropriately diluted unknown sugar sample solution, add 2 mL of DNS reagent and vortex.

(ii) Heat the tubes in a boiling water bath for 5 minutes, while the tubes are still warm, add 1 mL of 40% potassium-sodium tartrate solution. Cool the tubes to room temperature and add 7 mL of distilled water and vortex.

(iii) Measure the absorbance against the reagent blank at 540 nm. Construct a calibration curve on a graph paper, by plotting the sugar concentration

(0.2- 2 mg) on *x*- axis and absorbance at 540 nm on the *y*- axis. Compute the concentration of the sugar in the sample from the calibration curve. While calculating the sugar concentration in the unknown sample, the dilution factor has to be taken into account.

Protocol:

S.No	Standard maltose (mL)	Conc. of maltose (mg)	Distilled water (mL)	DNS reagent (mL)	Pot. Sod. tartrate (40%) (mL)	Distilled water (mL)	Absorbance at 540nm
1	Blank	- -	1.0	2.0	1.0	7.0	
2	0.1	0.2	0.9	"	"	"	
3	0.2	0.4	0.8	"	"	"	
4	0.3	0.6	0.7	"	"	"	
5	0.4	0.8	0.6	"	"	"	
6	0.5	1.0	0.5	"	"	"	
7	0.6	1.2	0.4	"	"	"	
8	0.7	1.4	0.3	"	"	"	
9	0.8	1.6	0.2	"	"	"	
10	0.9	1.8	0.1	"	"	"	
11	1.0	2.0	- -	"	"	"	
12	Sample-1	(to be determined)	"	"	"		
13	Sample -2	(to be determined)	"	"	"		

" = Same volume

Workout: Estimate the total reducing sugars in apple juice.

§ Miller, G. I. Anal. Chem.. 31: 426, (1972).

Experiment #21: **Determination of fructose by *Roe's* resorcinol method**[§]

Principle: Ketosugars undergo rapid dehydration to hydroxymethyl furfural than aldosugars. The hydroxymethyl furfural formed, later condenses with resorcinol to give a pink coloured complex with an absorption maximum at 520 nm.

Standard fructose stock solution (1 mg/mL): (i) Weigh 100 mg of fructose and transfer it carefully into a 100 mL volumetric flask. Make up the volume to 100 mL with distilled water (ii) Working standard fructose solution (100 µg/mL): Dilute 10 mL of the stock solution to 100 mL with distilled water in a volumetric flask.

Reagents: (i) Resorcinol reagent: Resorcinol (0.1% w/v) and thiourea (0.25 % w/v) in glacial acetic acid (ii) Dilute HCl: Mix concentrated hydrochloric acid and distilled water in the ratio of 5:1.

Procedure

(i)   To 2 mL of the standard sugar solution containing 10 - 80 µg of fructose or appropriately diluted unknown sugar sample solution, add 1 mL of resorcinol reagent followed by 7 mL of dilute HCl and vortex.

(ii)   Heat the tubes in a water bath, maintained at 80°C for 10 min. Cool to room temperature and read the pink colour developed, against reagent blank at 520 nm, within 30 min, in a photometer/colorimeter and record the absorbance.

(iii)   Construct a calibration curve on a graph paper, by plotting the fructose concentration (10 - 80 µg) on *x*- axis and absorbance at 520 nm on the *y*- axis. Compute the concentration of the sugar in the sample from the calibration curve. While calculating the sugar concentration in the unknown sample, the dilution factor has to be taken into account.

Protocol

S.No	Standard Fructose (mL)	Distilled water (mL)	Conc. of fructose (μg)	Resorcinol Reagent (mL)	Dilute HCl (mL)	Absorbance at 520 nm
1	Blank	2.0	- -	1.0	7.0	
2	0.1	1.9	10	"	"	
3	0.2	1.8	20	"	"	
4	0.3	1.7	30	"	"	
5	0.4	1.6	40	"	"	
6	0.5	1.5	50	"	"	
7	0.6	1.4	60	"	"	
8	0.7	1.3	70	"	"	
9	0.8	1.2	80	"	"	
10	Sample 1		(to be determined)	"	"	
11	Sample 2		(to be determined)	"	"	

" = same volume

Workout: Estimate the fructose content in the commercial honey sample.

§ (i) Roe, J. H. J. Biol. Chem. 107: 15, (1934).
   (ii) Roe, J. H. *et al.*, J. Biol. Chem. 178: 839, (1949).

Nucleic acid analysis

Experiment #22: Determination of DNA by diphenylamine method §

Principle: The deoxyribose moiety of the purine nucleotides of DNA reacts under acidic conditions to form a reactive species namely β-hydroxylevulinaldehyde. This aldehyde condenses with diphenylamine to give a blue coloured complex with an absorption maxima at 595 nm.

DNA standard solution (250 µg/mL) : Calf thymus or sperm whale DNA is used as a reference standard. Weigh accurately 25 mg of DNA and transfer it to a clean and dry 100 mL volumetric flask. Dissolve the DNA and make up the volume to 100 mL with citrate-buffered saline.

Reagents: (i) Buffered saline: Sodium citrate buffer (150 mM, *p*H 7.0) containing 0.9% sodium chloride (w/v) (ii) Diphenylamine reagent: Dissolve 1 g of crystalline diphenylamine in 100 mL of glacial acetic acid (w/v) with constant stirring, followed by addition of 2.5 mL of concentrated sulphuric acid (this reagent should be prepared freshly).

Procedure

(i)   To 2 mL of the standard solution (containing 50 - 500 µg of DNA) or undiluted/appropriately diluted unknown sample solution, add 4 mL of diphenylamine reagent and vortex (add the diphenylamine reagent using a pro-pipette or a burette). Heat the tubes covered with clean glass marbles in a boiling water bath for 10 min. Cool the tubes to room temperature and read the blue colour developed, against reagent blank at 595 nm, in a photometer/ colorimeter and record the absorbance.

(ii)   Construct a calibration curve on a graph paper, by plotting the concentration of DNA (50 - 500 µg) on *x*- axis and absorbance at 595 nm on the *y*- axis. Compute the concentration of the DNA in the sample from the calibration curve.

Protocol:

S.No	Standard DNA (mL)	Distilled water (mL)	Conc. of DNA (µg)	Diphenylamine reagent (mL)	Absorbance at 595 nm
1	Blank	2.0	- -	4.0	
2	0.2	1.8	50	"	
3	0.4	1.6	100	"	
4	0.6	1.4	150	"	
5	0.8	1.2	200	"	

6	1.0	1.0	250	"	
7	1.2	0.8	300	"	
8	1.4	0.6	350	"	
9	1.6	0.4	400	"	
10	1.8	0.2	450	"	
11	2.0	- -	500	"	
12	Sample - 1		(to be determined)	"	
13	Sample - 2		(to be determined)	"	

Workout: Estimate the content of DNA isolated from onion sample *(See, 9. Biochemical Preparations)*.

§ (i) Dische, Z. Mikro. Chemie. 8: 4, (1930).
  (ii) Grossman, L. & Moldave, K. (eds). *Methods in Enzymology*. Vol. XII. Nucleic Acids-Part B. Academic Press. NY, (1968).

Experiment #23 : Determination of RNA by orcinol method §

Principle: The ribose sugar moiety of RNA is converted to furfural under acidic condition. This derivative in the presence of ferric chloride reacts with orcinol to give a green coloured complex with an absorption maxima at 665 nm. This reaction is true for purine nucleotides only.

D-Ribose   Furfural   Orcinol   Green condensation adducts

RNA standard solution (100 µg/mL) : Yeast RNA is used as a reference standard. Weigh accurately 10 mg of RNA and transfer it to a clean and dry 100 mL volumetric flask. Dissolve the RNA and make up the volume to 100 mL with citrate-buffered saline.

Reagents: (i) Buffered saline: Sodium citrate buffer (150 mM, pH 7.0) containing 0.9 % sodium chloride (w/v) (ii) Orcinol reagent: Dissolve 0.1 g of ferric chloride in 100 mL of concentrated hydrochloric acid (w/v) and add 3.5 mL of 6 % (w/v) orcinol, dissolved in absolute ethanol. Store the reagent in a brown bottle.

Procedure

(i) To 3 mL of the standard solution (containing 10 - 150 µg of RNA) or undiluted/ appropriately diluted unknown sample solution, add 3 mL of orcinol reagent and vortex (add orcinol reagent using a pro-pipette or a burette). Heat the tubes covered with clean glass marbles in a boiling water bath for 30 min. Cool the tubes to room temperature and read the green colour developed, against reagent blank at 665 nm, in a photometer/ colorimeter and record the absorbance.

(ii) Construct a calibration curve on a graph paper, by plotting the concentration of RNA (10 - 150 µg) on x - axis and absorbance at 665 nm on the y-axis. Compute the concentration of the RNA in the sample from the calibration curve.

Protocol :

S.No	Standard RNA (mL)	Distilled water (mL)	Conc. of RNA (µg)	Orcinol reagent (mL)	Absorbance at 665 nm
1	Blank	3.0	- -	3.0	
2	0.1	2.9	10	,,	
3	0.2	2.8	20	,,	
4	0.3	2.7	30	,,	
5	0.4	2.6	40	,,	
6	0.5	2.5	50	,,	
7	0.6	2.4	60	,,	
8	0.7	2.3	70	,,	
9	0.8	2.2	80	,,	
10	1.0	2.0	100	,,	

11	1.2	1.8	120	"	
12	1.5	1.5	150	"	
12	Sample - 1	- -	(to be determined)	"	
13	Sample - 2	- -	(to be determined)	"	

" = Same volume

Workout : Estimate the RNA content in the commercial yeast tablets.

§ (i) Dische, Z. & Schwartz, K. Mikro. Chim. Acta. 2: 13, (1937).
  (ii) Grossman, L. & Moldave, K. (eds). *Methods in Enzymology.* Vol. XII. Nucleic Acids-Part B. Academic Press. NY, (1968).

Analysis of inorganic and total phosphorous

Experiment #24: Estimation of inorganic phosphorus by *Fiske-Subbarow* method §

Principle : Inorganic phosphate reacts with ammonium molybdate under acidic condition to form phosphomolybdic acid. Addition of a reducing agent (1-amino- 2-naphthol 4-sulphonic acid) reduces the molybdenum in the phosphomolybdate to give a blue coloured complex, the intensity of which is proportional to the amount of phosphate present. The blue coloured complex has an absorption maxima at 660nm.

Standard stock solution (13.6 mg/mL) : (i) Dissolve analytically pure KH_2PO_4 (1.36 g, mono-basic salt) in 30 mL of distilled water and make up the volume to 100 mL in a volumetric flask with distilled water. Add few drops of chloroform which acts as a preservative. (ii) working standard: Dilute, 1.0 mL of stock solution to 100 mL with distilled water to give a phosphorus concentration of 31 µg/mL.

Reagents : (i) 2.5 % (w/v) ammonium molybdate solution (ii) 5 N H_2SO_4 (iii) Amino naphthol sulphonic acid (ANSA) reagent: The powdered reagent is prepared by mixing 0.2 g of 1- amino - 2 - naphthol 4-sulphonic acid, 1.2 g of sodium bisulphite and 1.2 g of sodium sulphite. This mixture is ground to a fine powder using a porcelain mortar and pestle. Weigh 0.25 g of this powder and dissolve in 10 mL of distilled water (this solution should be prepared freshly).

Procedure

(i) To 1.0 mL of standard solution (containing 3.1 – 31 µg of inorganic phosphorus), add 1.0 mL of 5 N sulphuric acid, 1.0 mL of ammonium molybdate and 0.1 mL of reducing reagent (ANSA) and vortex. Make up the final volume to 10 mL by adding distilled water. Vortex and record the absorbance at 660 nm within 10 min against a reagent blank.

(ii) Construct a calibration curve on a graph paper, by plotting the concentration of inorganic phosphorus (3.1 - 31 µg) on *x*- axis and absorbance at 660 nm on the *y* -axis. Compute the concentration of the phosphorus in the sample from the calibration curve.

Protocol :

S.No	Standard phos-phorus (mL)	Distilled water (mL)	Conc. of phos-phorus (µg)	5NSul-phuric acid (mL)	Ammo-nium molybdate (mL)	ANSA reagent (mL)	Distilled water (mL)	Absor-bance at 660nm
1	Blank	1.0	- -	1.0	1.0	0.1	6.9	
2	0.1	0.9	3.1	”	”	”	”	
3	0.2	0.8	6.2	”	”	”	”	
4	0.3	0.7	9.3	”	”	”	”	
5	0.4	0.6	12.4	”	”	”	”	
6	0.5	0.5	15.5	”	”	”	”	
7	0.6	0.4	18.6	”	”	”	”	
8	0.7	0.3	21.7	”	”	”	”	
9	0.8	0.2	24.8	”	”	”	”	
10	0.9	0.1	27.9	”	”	”	”	
11	1.0	- -	31.0	”	”	”	”	
12	Sample - 1		(to be determined)	”	”	”	”	
13	Sample - 2		(to be determined)	”	”	”	”	

” = same volume

Workout: Estimate the amount of inorganic phosphorus and total phosphorus in bovine or human serum.

Method for sample processing: (a) *Inorganic phosphorus*:

Biological samples need to be deproteinized before the analysis. To 2 mL of serum sample add 8 mL of 10% trichloroacetic acid and vortex. Centrifuge at 4000 rpm for 10 min. or filter, use the supernatant for inorganic phosphorus estimation.

Protocol

Biological Sample (TCA supernatant) (mL)	5N H_2SO_4 (mL)	HNO_3	H_2O (mL)	Ammonium molybdate (mL)	ANSA reagent (mL)	H_2O (mL)	A_{660nm}
0.2	1.0	1 drop	0.8	1.0	0.1	6.9	
0.4	1.0	1 drop	0.6	1.0	0.1	6.9	

(b) *Total phosphorus* (*organic + inorganic phosphorus*).

After the addition of 5 N H_2SO_4, to the TCA supernatant, the test sample is evaporated in the test tube over a flame till it turn brown. On cooling, add one drop of 2 N HNO_3 and the heating is continued till the white flumes appear. If the contents do not turn colourless, 2 N HNO_3 addition step is repeated. After cooling, 1 mL of water is added and the tube is placed in a boiling water bath for 10 minutes. After cooling, 1 mL of ammonium molybdate and 0.1 mL of ANSA are added and the volume is made up to 10 mL with water. Record the absorbance at 660 nm.

The concentration of organic phosphorus is given by *(total phosphorus - inorganic phosphorus)*.

[§] Fiske, C. H. & Subbarow, Y. J. Biol. Chem. 66: 375, (1925).

Additional Reading

1.   Vogel, A. I. Text of Quantitative Analysis. 4th ed. ELBS, London, UK, (1982).

2.   Day, R. A. Jr & Underwood, A. L. Quantitative Analysis. 6th ed. Prentice-Hall of India, New Delhi, (1993).

3.   Holme, D. J. & Peck, H. Analytical Biochemistry. 3rd ed. Addison Wesley Longman Ltd. Essex, UK, (1998).

4.   Varley, H. Gowenlock, A. H. & Bell, M. Practical Clinical Biochemistry. 5th ed. William Heinemann Medical Books Ltd. London. (1980).

Lab Notes

Lab Notes

8

Biochemical Separation Techniques

Separation, identification and purity are important criteria in the analysis of biomolecules in order to assess their physico-chemical properties, biological activity and structural details. This is possible by the application of a variety of time honoured biochemical separation techniques. The routinely used biochemical methods of separation are essentially based on physical principles, such as solubility, density, molecular mass, conformation, charge, diffusion, adsorption, viscosity, thermal stability, chirality and biological specificity of biomolecules. Thus, it is possible to purify and characterize molecules of biological interest by application of a single biochemical technique or a combination of techniques, governed by different physical principles. It turns out that for biomolecules of natural origin, the later is true, as the desired molecule to be purified and analysed exists in a *milieu* of complex environment. The method(s) adopted during the course of the separation and purification of biomolecules should not be drastic, so that the native properties of the molecules are not destroyed (eg. biomacromolecules such as proteins and nucleic acids). Further, the selectivity of the method applied is important in order to achieve effective resolution of biomolecules. The resolution of a selected method(s) will depend on the differences in the physical property of desired and undesired molecules to be separated. For, instance a heterogeneous mixture of biomolecules with similar net charges or molecular mass cannot be separated based on the principles of ion-exchange or gel-filtration.

Amongst, various biochemical separation methods developed, chromatographic techniques such as partition, adsorption, ion-exchange, gel-filtration and affinity have been widely applied and exploited for separation and purification of biomolecules. In addition, electrophoretic techniques like, paper, polyacrylamide and agarose gel based methods are valuable tools in biochemical analysis. Based on these established principles, innovative and sophisticated methods such as flash chromatography, high performance liquid chromatography (HPLC), high performance thin layer chromatography (HPTLC), fast protein liquid chromatography (FPLC), isoelectric focusing, capillary and pulse field gel electrophoresis etc., have been developed for characterizing biomolecules with high resolution, sensitivity, and enhanced speed of performance.

Centrifugation, a separation technique based on molecular mass, density and shape is an important biochemical tool, widely used by biochemists for separation and characterization of macromolecules, cellular organelles and cell types. Further, spin columns, based on molecular weight cut off membranes with defined pore size have been successfully used for separation of desired molecules under applied centrifugal force.

In this chapter, experiments on chromatography, electrophoresis and dialysis are described. Chromatographic methods include (i) partition (ii) adsorption (iii) ion-exchange and (iv) gel-filtration techniques.

Experiment #1: Paper chromatography - separation of **(a)** Amino acids **(b)** Sugars **(c)** Purines & Pyrimidines.

(A) SEPARATION OF AMINO ACIDS

Principle: Amino acids are separated based on their partition or distribution coefficients, between the liquid stationary phase (water held by the chromatographic paper) and the liquid mobile phase.

Materials: (i) Chromatographic chamber (ii) Whatman No.1 filter paper (iii) Petri plate (iv) Capillary tubes (10 μL volume) (v) Hot air dryer (vi) Glass sprayer (vii) Enamelled tray (viii) Disposable plastic gloves.

Solvents system (mobile phase): *n*-Butanol : Acetic acid : Water (3:1:1). Prepare the solvent system by mixing 60 mL of butanol, 20 mL of acetic acid and 20 mL of water in a 250 mL glass beaker.

Spray reagent: Dissolve, 200 mg of ninhydrin in 99 mL of acetone and 1 mL of acetic acid.

Standard amino acids: L-Leucine, L-alanine, L-proline. Prepare, 2mg/mL of the individual amino acid solution, using distilled water. In case of leucine warm the solution in a boiling water bath to ensure the complete dissolution of the amino acid.

Procedure

(i)   Take Whatman No.1 filter paper of the following dimension - 50 x 20 cm (L x B). Handle the paper with disposable gloves. Draw a pencil line across the paper leaving 5 cm from the edge of the paper. This denotes the origin. Label the amino acids and the unknown to be spotted at the origin using a pencil. Spot the amino acid solution with the help of a capillary tube (10 μL). The diameter of the sample spotted should not exceed 3 - 4 mm. After application, dry the spot using a hot air dryer. Fold the filter paper in the shape of a cylinder and tie the edges at least at three points with cotton thread, so that the folded paper can stand upright.

(ii) Place the paper in a pre-saturated chromatographic chamber containing the mobile phase taken in a Petri plate. The spotted end edge should be placed in such a manner that the sample spots are 2 cm above the solvent level. Run the chromatogram (ascending chromatography) overnight (16 -18 h). After the run, the paper is removed, the solvent front is marked using a pencil and later air dried at room temperature.

(iii) The paper is unfolded and the chromatogram is sprayed or dipped in ninhydrin reagent. The chromatogram is allowed to develop at room temperature. Alpha -amino acids show purple colour while imino acids (proline) gives yellow colour. Outline the spots with a pencil. Measure the distance in centimetres from the origin to the center of the outlined spot and calculate the R_f (relative front/mobility) value for the separated amino acids.

$$R_f = \frac{\text{Distance moved by the amino acid (cm)}}{\text{Distance moved by the mobile phase (cm)}}$$

(iv) The unknown amino acids are identified by comparing their R_f values with that of standard amino acids (figure 8.1).

Note: Ninhydrin reagent to be handled using disposable gloves.

Workout: Identify some of the free amino acids in a sample of casein

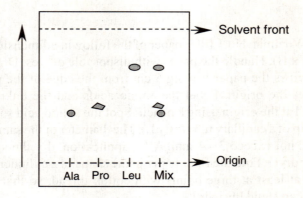

Fig. 8.1 A typical chromatogram of amino acids in B:A:W solvent system

hydrolysate (commercially available as a source of nitrogen for microbial growth medium), using standard reference amino acids.

(B) SEPARATION OF SUGARS

Materials: (i) Chromatographic chamber (ii) Whatman No.1 filter paper (50 x 20 cm) (iii) Petri plate (iv) Capillary tubes (10 µL volume) (v) Hot air dryer (vi) Glass sprayer (vii) Enamelled tray (viii) Disposable plastic gloves.

Solvents system (homogeneous mobile phase): Pyridine : Ethyl acetate : Acetic acid: Water (5:5:3:1). Prepare the solvent system by mixing 50 mL of Pyridine, 50 mL of ethyl acetate, 30mL of acetic acid and 10 mL of distilled water in a 250 mL glass beaker.

Spray reagent:(Aniline hydrogen phthalate reagent): Dissolve, 1g of distilled aniline and 1.6 g of phthalic acid in 500 mL of methanol.

Standard sugars: D-glucose, D-xylose, D-lactose. Prepare, 2mg/mL of the individual sugar solution.

Procedure:
(i)   Take Whatman No.1 filter paper of the following dimension - 50 cm

x 20 cm. Handle the paper with disposable gloves. Draw a pencil line across the paper leaving 5 cm from the edge of the paper. This denotes the origin. Label the sugar and the sugar mixture or test samples to be spotted at the origin, using a pencil. Spot the sugar solution with the help of a capillary tube (10 μL). The diameter of the sample spotted should not exceed 3 - 4 mm. After application, dry the spot using a hot air dryer. Fold the filter paper in the shape of a cylinder and tie the edges at least at three points with cotton thread, so that the folded paper can stand upright.

(ii) Place the paper in a pre-saturated chromatographic chamber containing the mobile phase taken in a Petri plate. The spotted edge should be placed in such a manner, that the sample spots are 2 cm above the solvent level. Run the chromatogram (ascending chromatography) overnight (16 -18h). After the run, the paper is removed, the solvent front is marked using a pencil and later air dried at room temperature.

(iii) The paper is unfolded and the chromatogram is sprayed with or dipped in aniline hydrogen phthalate reagent (using enamelled tray). The chromatogram is dried at room temperature and placed in a hot air oven maintained at 110°C for 10 min for colour development. Hexose sugars appear as brown coloured spots, while pentose sugars form pink spots. Disaccharides react slowly as compared to monosaccharides with the spray reagent.

(iv) Record, calculate and compare the R_f values of the standard and the test sample or sugar mixture.

Note: Use distilled pyridine or Analytical grade/GR solvent.

Workout: Identify the sugars present in commercial apple juice.

(C) PURINES & PYRIMIDINES

Materials: (i) Chromatographic chamber (ii) Whatman No.1 filter paper (iii) Petri plate (iv) Capillary tubes (10 μL volume) (v) Hot air dryer (vi) UV lamp (shortwave - 260-300 nm wavelength radiation) or UV cabinet.

Solvent system (mobile phase - isopropanol- HCl): Mix 65 mL of isopropanol, 16.7 mL of Conc. hydrochloric acid and make up the volume to 100 mL with distilled water.

Standard purines & pyrimidines: Adenine, guanine, cytosine and uracil (2 mg/mL). Prepare the standards in 0.1 N HCl (warm the solution if necessary for dissolving).

Procedure

(i)   Take Whatman No.1 filter paper of the following dimension - 40 x 15 cm (L x B). Handle the paper with disposable gloves. Draw a pencil line across the paper leaving 5 cm from the edge of the paper. This denotes the origin. Label the purine/pyrimidine and the mixture of free bases or test samples to be spotted at the origin, using a pencil. Spot the standards and the sample solution with the help of a capillary tube (10 μL).The diameter of the sample spotted should not exceed 3 - 4 mm. After application, dry the spot using a hot air dryer. Fold the filter paper in the shape of a cylinder and tie the edges at least at three points with cotton thread, so that the folded paper can stand upright.

(ii)   Place the paper in a pre-saturated chromatographic chamber containing the mobile phase taken in a Petri plate. The spotted edge should be placed in such a manner, that the sample spots should be 2 cm above the solvent level. Run the chromatogram (ascending chromatography) overnight (16 -18h). After the run, the paper is removed, the solvent front is marked using a pencil and later air dried at room temperature.

(iii) The air dried paper is unfolded and the separated bases on the chromatogram are visualized as fluorescent spots under a UV lamp, in a dark room. These spots are outlined with a pencil.

       Caution: Do not view UV lamp directly. Wear UV protective goggles while viewing.

(iv)   Record, calculate and compare the R_f values of the standards and the test sample or base mixture.

Workout: Elute one of the separated bases in 0.1 N HCl and record the UV spectra.

Experiment #2: **Separation of amino acids by paper electrophoresis**

Principle: The separation of charged biomolecules under the influence of electrical field forms the basis for electrophoresis. The charge therein is dependent on the *p*H of the medium. At *p*H 5.5, the basic amino acids (arginine, lysine) exist as cations, neutral amino acids (alanine, leucine etc.,) as zwitterions and acidic amino acids (aspartic acid, glutamic acid) as anions. Separation of amino acid mixture can be achieved by paper electrophoresis.

Materials: (i) Electrophoretic apparatus (horizontal or vertical type unit) (ii) Power pack (iii) Whatman No. 1 filter paper (iv) Glass sprayer (v) Petri plate (vi) Capillary tube/ micro-syringe (10 µL) (v) Green gram.

Reagents: (i) Sodium acetate buffer 50 mM (*p*H 5.5) (ii) Spray reagent: Ninhydrin (0.2% w/v, in 99 mL of acetone and 1 mL of acetic acid) (iii) Trichloroacetic acid (10%w/v).

Standard amino acids: L- arginine, L-aspartic acid, L-leucine (1 mg/mL). Prepare 1 mg/mL of the amino acid in distilled water. In case of aspartic acid, the solubilization of the amino acid is effected by addition of small volumes of ammonium hydroxide.

Procedure

(i)   Cut Whatman No.1 filter paper strips (size: 160 mm x 20 mm) or appropriate dimension depending upon the size of the electrophoretic apparatus. Draw a pencil line at the center of the strip (denotes origin). Indicate the (+) and (-) symbol at the end of the paper strip. Apply, 10 µL of the individual amino acids or the test sample at the origin as a streak, using a capillary tube/micro-syringe. After application, dry the strip using a hot air dryer.

(ii)   Fill the cathodic (-) and anodic (+) chambers of the electrophoretic apparatus with equal volumes of acetate buffer. Place the paper strips in the electrophoretic apparatus, with the (+) and (-) ends of the strip dipping into anodic (colour code; black) and cathodic (colour code: red), respectively. Moisten the paper strip with acetate buffer, using a 1 mL glass pipette, leaving a distance of 2 cm from the origin. Allow the unmoistened region of the paper strip to wet by capillary action.

Close the lid of the electrophoretic apparatus and connect the respective terminals to the power pack. Switch on the power pack and run the electrophoresis at 400 V for 70 -100 min.

(iii)   After the run, the power supply to the apparatus is switched off. The paper strips are removed and air dried at room temperature. Dip the paper strips in ninhydrin reagent taken in a Petri plate or spray the ninhydrin reagent onto the strip in a fume hood. Allow the strips to develop at room temperature. The separated amino acids appear as purple bands. Basic and acidic amino acids show cathodic and anodic mobilities, respectively, while, the neutral amino acids remain at the origin. However, it should be noted that the neutral amino acids exhibit slight cathodic mobility due to electro-osmosis.

(iv)   Mark the separated amino acid bands with a pencil. Measure the electrophoretic mobility of the amino acid from the origin, in centimetres. The identification of the amino acid(s) in the test sample is achieved by comparing the electrophoretic mobilities of the standard amino acids (figure 8.2).

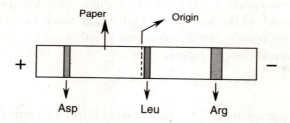

Fig.8.2 Diagrammatic representation of separation of amino acids by paper electrophoresis

Workout: Analyse the amino acid profile of germinated green gram (*Phaseolus aureus* Roxb.)

Procedure: Take 5 g of green gram (whole seeds) and soak it in 15 - 20 mL of tap water, over night, at 37°C. Place the pre-soaked seeds in a sterile glass Petri-plate containing moist cotton. Allow the seeds to germinate over night. Homogenize the germinated seeds (1 gm) in 5 mL of distilled water. Take 1 mL of the homogenate and add an equal volume 10% TCA. Vortex and centrifuge the sample at 3000 rpm, for 20 min. Transfer 0.7 mL of the clear supernatant into a test tube and remove the TCA by repeated extraction with 3 mL portions of ether. Examine the lower aqueous phase for the presence of free amino acids.

Experiment #3: Determination of N-terminal amino acid of a dipeptide by fluorodinitrobenzene (FDNB) method [s]

Principle: Fluorodinitrobenzene (Sanger's reagent) reacts with the primary amino group of the N-terminal amino acid and with phenolic, thiol, imida-

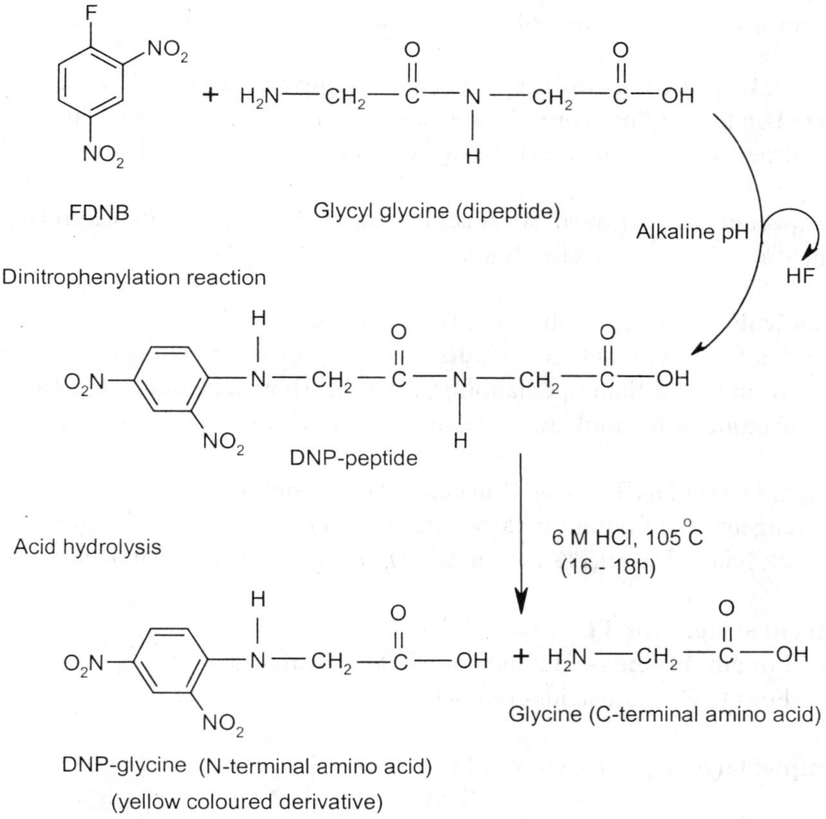

FDNB   Glycyl glycine (dipeptide)   Alkaline pH   HF

Dinitrophenylation reaction

DNP-peptide

Acid hydrolysis   6 M HCl, 105 °C (16 - 18h)

DNP-glycine (N-terminal amino acid)   Glycine (C-terminal amino acid)
(yellow coloured derivative)

DNP - Dinitrophenyl

N-terminal amino acid determination of glycyl glycine by Sanger's method

zole and ε-amino groups of the peptide and protein, to yield the dinitrophenylated (DNP) peptide or protein derivative, at alkaline *p*H. Acid hydrolysis of the DNP-derivative results in the release of the N-terminal amino acid as its DNP-derivative (yellow coloured derivative); While the rest of the amino acids which are in peptide linkage are converted either to free amino acids or DNP-amino acids, formed due to reactive groups, other than the N-terminal amino acid. The identity of the DNP-amino acid is established by application of chromatographic techniques, such as paper, TLC, HPLC etc., along with authentic DNP-amino acids.

Note: Majority of the DNP-amino acids are ether soluble, while some are water soluble (ex. DNP-arginine, O-DNP-tyrosine, ε-DNP-lysine and imidazole-DNP-histidine).

Materials: Pre-coated-Silica gel G and Cellulose plates (size 20 x 20 cm; Coated on to aluminum or polyester sheets), glass hydrolysis vial (ii) Bunsen burner (iii) Pro-pipette (iv) TLC developing tank with lid.

Equipment: Hot air oven, wrist action shaker, hot water bath/steam bath, Centrifuge, UV-Vis spectrophotometer.

Chemicals: Fluorodinitrobenzene (FDNB) (**Caution**: Corrosive chemical), peroxide free diethyl ether* (**Caution**: Highly inflammable solvent; keep away from naked flame), ethanol, sodium bicarbonate, Conc. HCl, ninhydrin, acetone, *n*-butanol, acetic acid, benzyl alcohol.

Reagents: (i) FDNB (5% w/v) in ethanol (**Caution**: No mouth pipetting of this reagent) (ii) Sodium bicarbonate solution (0.4%w/v) (iii) Ninhydrin reagent: Ninhydrin (0.2% w/v) in 99 mL acetone and 1 mL of acetic acid.

Solvent system for TLC:
(1)   For amino acids – Butanol:Acetic acid:Water (3:1:1).
(2)   For DNP-amino acids- Benzyl alcohol:Ethanol (9:1).

Sample: Glycyl glycine (MW 132).

Reference DNP-amino acid: DNP-glycine.

---

* Peroxide free ether is prepared by adding 3 - 5 g of ferrous sulphate and 15 – 20 mL of distilled water to 1L of diethyl ether (water saturate ether). Store the solvent in a chemical safety hood.

Procedure

(i)   Weigh and transfer, 2 mg (15 µ moles) of glycyl glycine into a stop-pered graduated conical test tube (15 mL capacity). Dissolve the dipep-tide in 0.5 mL of sodium bicarbonate solution (0.4% w/v). Add, twice the volume of FDNB reagent (1 ml) and shake the contents gently at ambient temperature for 2h using a wrist action shaker. Monitor the pH, of the reaction from time to time, using a narrow range pH indi-cator paper, so as to maintain an alkaline condition (pH 8-9).

(ii)   After 2h, unreacted FDNB in the reaction mixture is removed by re-peated extraction with 3 mL peroxide free ether. Continue the extrac-tion until the ether extract is colourless (discard the ether extracts). Acidify the lower aqueous phase by adding a drop of conc. HCl (25-50 µL) and vortex the contents. (**Caution**: Carry out the ether extrac-tions in a fume hood).

(iii)   Extract the DNP-peptide from the acidified aqueous phase with small volumes of ether (2 mL per extract), until the ether extract is colourless. Pool the ether extracts into a clean glass beaker. Evaporate the ether gently in small aliquots over a steam bath to dryness, inside a fume hood. Dissolve the dried yellow residue (ether free) in 0.5 -1 mL of acetone. Transfer the contents to a glass hydrolysis vial. Evaporate the acetone by placing the tube in a steam bath. Add 0.5 mL of 6 M HCl to the residue and seal the vial over a Bunsen flame. Subject the sample to acid hydrolysis by placing the sealed vial in a hot air oven, maintained at 105°C for 18 hours.

(iv)   After hydrolysis, remove the sealed vial and bring the contents to room temperature. Break the seal cautiously, transfer the contents into a test tube and add 2.0 mL of distilled water. Extract the acid hydroly-sate with small volumes (2 mL) of ether. Repeat the extraction, until the ether extract is colourless. Pool and transfer the ether extracts into a test tube. Evaporate the ether extract to dryness in small aliquots, over a steam bath. Redissolve the yellow residue obtained, in mini-mum volume of acetone (≤ 0.3 mL). Use the aqueous phase for iden-tification of the C-terminal amino acid of the dipeptide.

(v) Spot 5 - 10 µL of the N-terminal DNP-amino acid present in acetone on to a pre-coated silica gel-G TLC plate, along with authentic DNP-glycine. Develop the chromatogram in Benzyl alcohol:Ethanol (9:1) solvent system. Identify the yellow coloured N-terminal amino acid, by comparing the R_f values.

Note: (i) Invariably an additional yellow spot of dinitrophenol, as a contaminant is observed on the chromatogram, which can be identified by its decolorization upon exposure to HCl fumes (carry out this operation in a fume hood); while the colour of the DNP-amino acid remains unchanged.

(ii) **Preparation of DNP-glycine**: Weigh and transfer 20 mg (0.27 mmole) glycine and 65 mg (0.77 mmole) of sodium bicarbonate into a test tube. Dissolve them in 0.5 mL of distilled water. Add 1 mL of 4% (w/v) FDNB in ethanol. The reactants are mixed for 2h at room temperature, using a wrist action shaker. Excess, FDNB is removed from the reaction mixture by repeated extraction with peroxide free ether. Later the DNP-glycine is precipitated by acidifying the aqueous phase with 1N HCl. Separate the yellow precipitate by centrifugation at 2500 rpm, for 10 min. Discard the supernatant and crystallize the DNP-glycine from aqueous methanol. Use this derivative as the reference, in case commercial standard is not available.

Workout: (i) Determine the C-terminal amino acid of the dipeptide by evaporating the aqueous phase to dryness, followed by its TLC analysis on a precoated cellulose plate, along with authentic glycine, as reference. Use solvent system -1, for chromatography. Identify the amino acid by ninhydrin reagent. (ii) Determine the absorption maxima (λ_{max}) of DNP-glycine in 1% sodium bicarbonate solution, in a UV-Vis spectrophotometer.

[§] David, G (ed). *Methods of Biochemical Analysis*. Vol. II. Interscience Publishers, Inc. NY, USA, (1961).

Experiment #4: **Separation of small and large molecules by dialysis**

Principle: The gross separation of small molecules from larger ones by diffusion through a semipermeable membrane is dialysis. The rate of diffusion is enhanced by increasing temperature. However, this is not applicable to proteins as they tend to denature.

Materials:(i) Dialysis tubing (Average flat width 25 mm; dry cylinder diameter 16 mm; molecular weight cut off - ≥ 12 kDa (ii) Magnetic stirrer and magnetic bead.(iii) plastic clamps.

Reagent: Barium chloride solution (5%w/v).

Sample: A solution of bovine serum albumin (10 mg/mL) and ammonium sulphate (50 mg/mL). [MW of (i) BSA - 66 kDa, and (ii) ammonium sulphate - 132}.

Procedure

(i)   Take 10 cm long dialysis tubing. After washing the dialysis tubing, knot at one end or use a dialysis plastic clamp. Transfer 5 mL of protein-salt solution using a pipette and clamp/knot the other end of the dialysis tube. Suspend the tube in a 2 L glass beaker containing distilled water. The contents are gently stirred using a magnetic stirrer in a refrigerator at 4°C, overnight.

(ii)   Next day remove the water and test the water for the presence of sulphate ions by addition of barium chloride solution (presence of sulphate is indicated by appearance of white turbidity/precipitate of barium sulphate). Add fresh distilled water and continue the dialysis until the distilled water tests negative for sulphate. This indicates the complete separation of salt from protein solution.

Workout

(i)   Determine the protein content of the dialysed sample by Lowry method and calculate the % recovery of protein.

(ii)   Separate a mixture of blue-dextran (2 mg mL^{-1} in distilled water) and potassium dichromate (10 mg mL^{-1} in distilled water) solution by dialysis. Compare the observation with gel-filtration technique.

Note: Transfer the dialysed protein solution into a measuring cylinder (25 mL). Rinse the tube with distilled water so as to recover the protein completely. Note the volume before the protein estimation.

Experiment #5: Separation of blue-dextran and potassium dichromate by gel-filtration chromatography.

Principle: Gel-filtration, also known as molecular sieve or exclusion-chromatography is a technique employed in the separation of biomolecules.

Separation of substances according to their molecular sizes, using porous gel beads forms the basis of gel-filtration technique. A solution containing analytes of different sizes is allowed to pass through the gel matrix. The solute molecules, whose molecular dimensions are larger than the pores, cannot enter gel matrix, hence their accessibility is restricted to the inter-spaces (void space) in between the gel beads, and thus the larger molecules are excluded out rapidly. On the other hand, the small molecules enter into the porous interior of the gel bead and are capable of diffusing in and out of the gel beads, i.e., a larger volume of the solvent is required for their elution. Thus, the mode of elution in gel-filtration is directly related to the molecular dimensions of the analyte. However, the secret of success in gel-filtration is mainly governed by porosity of the gel bead and additional factors such as flow rates, temperature and nature of solvent are equally important (figure 8.3).

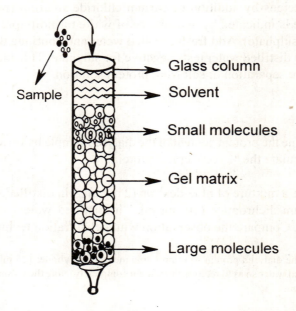

Fig. 8.3. Gel-filtration Chromatography

Materials: Glass column with a tapered end [10 cm x 1.5 cm (dia)] or 10 mL plastic disposable syringe, Sephadex G-10, Blue-dextran, potassium dichromate, siliconized glass wool or glass wool, screw clip, silicone tube, burette stand with clamp.

Sample: Prepare a mixture of blue-dextran (MW 2000 kDa-high molecular weight polysaccharide) and potassium dichromate (MW 294 Da) in distilled water (concentration - 2 mg of each in 1 mL of distilled water).

Procedure

(i) Weigh 8-10 g of Sephadex G-10 into a 150 mL glass beaker, add 50 mL of distilled water and stir the contents. Allow the gel to swell for 3-4 hours. Decant the solution to remove the fines. Wash the gel suspension once again with distilled water. To regulate the flow rate, the tapered end of the column is fitted with a 4-5 cm long silicone tubing having a screw clip. Pour the gel into a glass or plastic column, plugged with glass wool at the bottom end, so as to hold the gel matrix from running out of the column. Allow the gel to settle under gravity. Equilibrate the column by passing 20 - 25 mL distilled water.

Note: (i) Avoid the formation of air bubbles while packing the gel matrix (ii) packing should be uniform without any banding (iii) Column matrix should never be allowed to dry.

(ii) Apply 0.3 mL of sample solution onto the surface, without disturbing the gel matrix. After the sample enters the gel matrix, elute the sample components with distilled water at a low flow rate (20 mL/h). Collect twenty, 1mL fractions in graduated plastic Eppendorf tubes. Blue-dextran does not enter the gel pores, hence it is excluded out earlier, while potassium dichromate is retarded by the gel matrix as they enter the gel pores, thus it elutes later. The separation of these molecules in the gel column can be monitored visually, as they are coloured. Record the absorbance of the fractions collected (at 630 nm for blue dextran and 340 nm for potassium dichromate). Plot an elution profile graph (*x*- axis - fraction number *vs y* axis – absorbance). Figure 8.4 gives a typical elution profile of separation of high and low molecular weight substances.

Workout: Separate a mixture of BSA and tryptophan (1 mg/mL in 0.01 N NaOH) using Sephadex G-10 column. Elute the loaded sample with 0.01 N NaOH. Measure the absorbance of the fractions collected at 280 nm, using a UV-Vis spectrophotometer. Draw an elution profile by plotting fraction number *vs* absorbance.

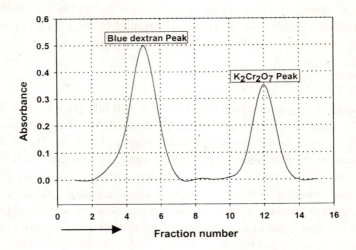

Fig. 8.4 Gel filtration (Sephadex G-10): Elution profile

Note: Equilibrate the gel-column with 0.01N NaOH solution, before application of the sample.

Experiment #6: Separation of amino acids by ion-exchange chromatography

Principle: The separation of charged molecules using an insoluble polyelectrolyte matrix forms the basis of ion-exchange chromatography. When a solution containing different charged species is allowed to pass through ion-exchange matrix (resin), the molecules irrespective of their sizes carrying charges opposite to that of ionizable groups of the exchanger, get attracted and bound due to electrostatic attraction. Molecules with similar charges will pass freely due to electrostatic repulsion. The bound molecule can be eluted from the resin by changing either the *p*H or the ionic strength of the eluent. The binding of the charged ions depends upon the type and number of ionizable groups on the ion exchange matrix, *p*H and ionic strength of the eluent. Depending on the charge of the molecule to be separated (-) or (+), one can employ either cation or anion exchanger in ion-exchange chromatography. Sulphonated polystyrene cross linked with divinyl benzene (ex. Dowex-50) is a strong cation exchange resin, which has been widely used in the separation of amino acids (figures 8.5 & 8.6).

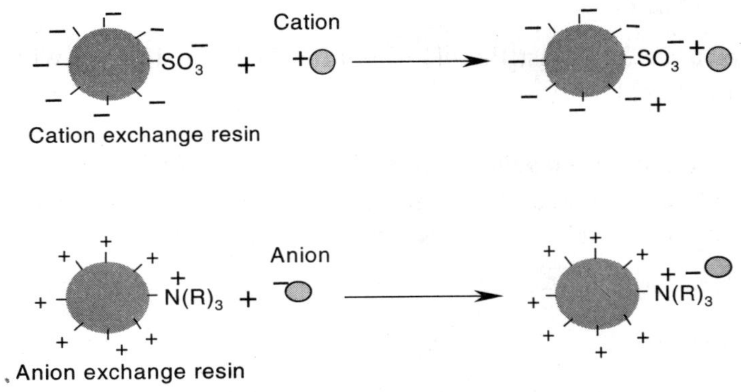

Fig. 8.5 Ionic interaction of ion-exchange resin.

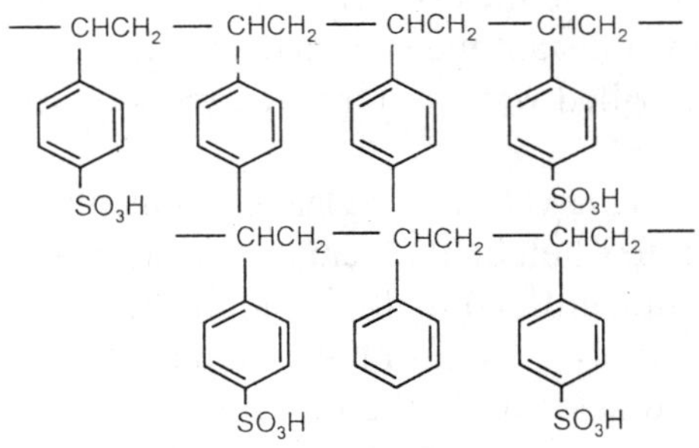

Fig. 8.6 Partial structure of sulphonated polystyrene resin
(Cation exchanger).

Materials: (i) *Amberlite* IR-120 [cation exchanger with sulphonic functional group] (ii) glass column with a tapered end (15 x 1 cm) connected with silicone tubing and screw clip (iii) burette stand and clamp (iv) siliconized glass wool (v) Amino acids: L- glutamic acid and L- arginine.

Reagents *for cation exchange chromatography with Amberlite IR-120 resin*: (i) Sodium citrate buffer, *p*H 6.0, 0.2 M (ii) Sodium phosphate buffer, *p*H 6.0, 0.2 M.

Elution buffer 1: Mix two volumes of citrate buffer with one volume of phosphate buffer.

Elution buffer 2: Adjust the pH of elution buffer-1, to 10.5 with sodium hydroxide (10 N).

Preparation of amino acid mixture (2mg/mL): Weigh 50 mg of each amino acid (L-glutamic acid and L- arginine) and transfer them into a clean and dry 25 mL volumetric flask. Dissolve and make up the volume with elution buffer-1.

Procedure

(i)   Take 25 g of *Amberlite* resin into a glass beaker and suspend it in 50 mL of 1 N NaOH. Stir the contents with a glass rod, allow the resin to settle and later decant the supernatant. The resin is now in Na$^+$ form and is repeatedly washed with distilled water. After the last wash, add 50 mL of 1 N HCl to the resin and stir the contents. Allow the resin to settle, and then decant the supernatant. The resin is washed several times with distilled water to remove excess of acid.

(ii)   Pack the *Amberlite*-(H$^+$) into a glass column, plugged with glass wool at the bottom to a height of 10 cm. Equilibrate the column by passing 50 mL of elution buffer-1, before loading the amino acid mixture onto the column. Apply 0.5 mL of t.ie amino acid mixture solution (1 mg of each amino acid) on to the top of the resin, by gently running it down. Rinse any remnants of the sample by adding 0.5 mL of elution buffer-1 and allow it to drain. Repeat the rinse step, once more and continue the elution with elution buffer-1, at a flow rate of 30 mL/ hour. Collect, ten - 3 mL fraction in the serially numbered test tubes. This is followed by elution with elution buffer-2, collecting ten, 3 mL fractions.

(iii) Analyse each fraction collected for the presence of amino acid by performing ninhydrin reaction (*see 7*, **Quantitative analysis**). Use 0.5 mL aliquot for analysis. Draw the amino acid elution profile by plotting fraction # vs absorbance at 570 nm.

Workout: Separate an amino acid mixture containing L-leucine and L-lysine (1mg each) by the above method. Compare the elution profile with that obtained by a mixture of L-glutamic acid and L- arginine.

Experiment #7: Separation of plant pigments by Thin layer chromatography (TLC).

Principle: The separation of compounds in TLC is based on differential adsorption as well as partitioning of the analytes between the liquid stationary phase and mobile solvent phase. This technique is rapid as compared to paper chromatography. Molecules get separated between the hydrated stationary phase and non-polar mobile phase. Hydrophilic analytes have more affinity to the polar stationary matrix, while less hydrophilic molecules tend to have more affinity towards mobile phase, resulting in its faster movement and separation. The separated analytes are identified by comparing their R_f values to that of reference standards. The R_f value of an analyte depends upon the (i) solvent system (ii) degree of saturation of the mobile phase in the chromatographic chamber (iii) particle size of the adsorbent (iv) type of adsorbent (v) temperature and humidity. Thus, the R_f value for an analyte is constant for a given set of experimental conditions.

$$R_f = \frac{\text{Distance travelled by the analyte}}{\text{Distance travelled by the mobile phase (solvent)}}$$

Commonly used stationary matrix for TLC include, silica gel - G, silica gel -H, micro-porous cellulose, alumina, florisil, polyamide and octadecylsilane. Normally, glass, aluminium or polyester supports (size 20 x 20 cm) are used for coating stationary matrixes.

Material: (i) Glass plates (20 x 20 cm, thickness 3 mm) (ii) Chromatographic glass jar with vacuum greased lid (iii) TLC applicator with plastic plate support and TLC rack (iv) TLC-spotting guide (v) Calibrated capillary tubes (5 µL/10 µL capacity) or micro-syringe (10 µL). Silica gel-G, TLC grade.

Sample: *Spirulina* powder (commercially available in drug stores/medical shops as capsules)

Solvent system: Petroleum ether: Acetone (7:3 v/v).

Procedure

(i)   Arrange 5 clean glass plates on the plastic support unit along with
        silica gel applicator. Weigh 40 g of silica gel-G into a 250 mL iodine
        flask. Add 100 mL of distilled water and thoroughly shake the contents.
        Immediately, pour the slurry into the applicator and coat the gel along
        the glass plate, by rapidly moving the applicator. The coating of the
        adsorbent should be of uniform thickness (25 - 75 μm). Air dry the
        plates (placed in TLC rack) at room temperature for 5 hours. Activate
        the coated TLC plates in a hot air oven at 110°C for 1 hour. Later, the
        plates are removed and allowed to cool at room temperature in a
        desiccated chamber.

(ii)   Prepare an acetone extract of the *Spirulina* powder (500 mg), by
        transferring the contents of one capsule into a 20 mL glass test tube
        containing 5 mL of acetone. Vortex and allow the contents to stand at
        room temperature for 30 minutes. Collect the acetone extract by filtering
        through glass wool. The clear filtrate is concentrated by evaporation
        to a small volume (500 μL volume).

(iii)   Spot equidistantly 5, 10, 15 and 20 μL aliquots of the acetone extract
        onto the activated silica gel plate, about 3 cm away from the edge of
        the plate by using a TLC guide and capillary tube. The area of the spot
        should be kept to minimum (1-2 mm dia), which can be achieved by
        repeated spotting of the sample volume followed by air drying at the
        same spot. The chromatogram is developed by placing the TLC plate
        vertically in a TLC chamber, saturated with the mobile phase, in such
        a manner that the spotted edge dips into the solvent system. Run the
        chromatogram until the solvent front reaches the top edge of the plate.
        Remove and mark the solvent front as soon as the plate is removed
        from the chamber. Air dry the plate at room temperature. Outline the
        coloured pigment spots using a poker and calculate the R_f values of
        the pigments.

        Caution: Do not touch or damage the coated area and the edges of
        the plate.

        For the purpose of identification, following are the reference R_f values
        (Table 8.1) of the pigments in the petroleum ether: acetone solvent
        system.

Table 8.1 R_f values of plant pigments

Pigment	R_f value
Chlorophyll *a*	0.68
Chlorophyll *b*	0.54
Chlorophyll *c*	0.03
β-carotene	0.94
Fucoxanthin	0.51
Lutein	0.43
Violaxanthin	0.22

Workout: Analyse the pigments present in *Spinach* leaves, by TLC and compare the profile with *Spirulina*.

Note: The separated pigment spots on the TLC plate should be scored immediately, as the pigment undergoes destruction/bleaching in the presence of light.

Additional Reading

1.   Boyer, R. F. Modern Experimental Biochemistry, 2nd ed. The Benjamin/Cummings Publishing Company, Inc. California, USA, (1993).

2.   Wilson. K. & Walker, J. Principles and Techniques of Practical Biochemistry. 4th ed. Cambridge University Press. Cambridge, UK, (1995).

3.   Cooper, T. C. Tools of Biochemistry, John Wiley, NY, USA, (1997).

4.   Hawcroft, D. M. Electrophoresis the Basics. IRL Press, Oxford, UK (1997).

5.   Holme, D. J. & Peck, H. Analytical Biochemistry. 3rd ed. Addison Wesley Longman Ltd. Essex, UK, (1998).

Lab Notes

Lab Notes

9

Biochemical Preparations

The art of obtaining a pure biomolecule of interest from its natural source by application of different biochemical techniques forms the basis of biochemical preparations. Purification, subsequent to extraction of the biomolecule from the natural source depends upon the successful removal of co-contaminants in the desired purified product. The experimental procedures adopted for the biochemical preparation deals with a variety of purification methods that remove the contaminants. The methodology may involve either a single step preparative procedure or a combination of procedures (multi-step). It is of prime importance that the methods applied during the course of isolation and purification should not result in the alteration of native biological properties of the purified molecule.

The other important criteria of biochemical preparations are yield and purity of the final desired product. Figure 9.1 gives the general outline involved in biochemical preparation. The techniques involved in isolation and purification of biomolecules include well established techniques encompassing (i) precipitation (isoelectric, salt, solvent and thermal denaturation), (ii) preparative centrifugation (differential and density gradient) and (iii) chromatographic techniques involving biospecificity (affinity), charge (ion-exchange) and size based (gel-filtration methods). The purity of the desired preparation are evaluated by chromatographic, electrophoretic and immunoanalytical methods.

One of the essential requirements for selecting a natural source for isolation of biomolecule of interest is its easy availability and inexpensive sourcing. Natural sources for biochemical preparation include, cells, tissues, eggs, biological fluids such as blood, serum plasma, amniotic fluid, urine, milk etc., plant exudates, plant roots, seeds and fruits. In this chapter, simple laboratory procedures for separation and purification of biomolecules (amino acids, proteins, carbohydrates, lipids and nucleic acids) both from animal and plant origin are described.

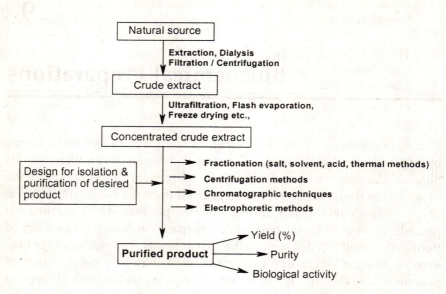

Fig.9.1 Flow chart outlining the general steps involved in biochemical preparations.

Experiment #1: Isolation of glutamic acid from gluten

Principle: Gluten, the major wheat protein is a rich source of glutamic acid. This amino acid contributes ~30% of the total amino acid content of the protein. Glutamic acid is isolated from gluten by acid hydrolysis followed by crystallization of glutamic acid.

Reagents: (i) Concentrated HCl (ii) Acetone (iii) Sodium hydroxide solution (4% w/v) (iv) Diethyl ether (v) Congo red indicator (0.1% w/v) in ethanol.

Sample: Whole wheat flour

Procedure

(i)   Weigh 50 g of wheat flour and prepare dough with addition of distilled water. Allow it to stand for 30 - 40 minutes in a closed container. Later, place the dough in a muslin cloth (use two folds of the cloth) and repeatedly dip and wash with distilled water to remove the starch. Squeeze out the residual water from the gluten and air dry the sample.

(ii) Transfer the sample into a 250 mL round bottomed glass flask and add 30 mL of concentrated HCl. Heat the contents on the steam water bath until the protein dissolves. Later, reflux the contents for 8 hours using a heating mantle and reflux condenser. To avoid bumping, add pumice stones or porcelain chips. Cool the contents to room temperature and dilute with equal volume of distilled water.

(iii) Concentrate the diluted protein hydrolysate to 10 - 15 mL by flash evaporation or over a steam bath using a China dish. Transfer the concentrate to a 150 mL conical flask and saturate the contents by bubbling with dry HCl gas. Cool and seed (pinch of pure glutamic acid) the contents and leave it in a refrigerator at 4°C for 2 - 3 days, so as to facilitate the process of crystallization of glutamic acid hydrochloride.

(iv) Filter the contents after addition of an equal volume of cold acetone using a sintered glass funnel/Büchner funnel. Wash the crystals with cold acetone followed by diethyl ether. Redissolve the dried sample in minimum volume of hot distilled water and neutralize the solution with drop wise addition of 4% NaOH, till it tests positive (use a small aliquot ~ 50 µL for the spot test on a white porcelain tile) with Congo red indicator to give a blue colour (*p*H 3.0 to 3.3). Centrifuge the sample at 3500 rpm for 10 min to remove any undissolved impurities. Concentrate the supernatant by flash evaporation to 5 mL. Allow the glutamic acid to crystallize by placing the sample in a refrigerator for 24 - 48 hours. Recover the crystals by filtration and air dry the sample. Record the weight and calculate % yield.

Workout: Analyse the purity of the preparation by paper chromatography using a reference standard.

Experiment #2: Preparation of cystine from sheep's wool or human hair

Principle: The protein keratin is a rich source for sulphur containing amino acids, such as cystine. The amino acid can be obtained from keratin by acid hydrolysis followed by its precipitation with sodium acetate solution, at pH 5.0.

Reagents/chemicals: (i) Formic acid (50% v/v) solution (ii) Sodium acetate (50 % w/v) solution (iii) Conc. hydrochloric acid (35-36%) (iv) Dilute HCl (3%) solution (v) Chilled diethyl ether (vi)Activated charcoal (vii) Narrow range pH (3 -5) indicator paper or Congo red pH indicator solution (viii) Ammonium hydroxide solution (ix) Ethanol.

Extraction reagent: Mix 100 mL of conc. HCl with 400 mL of 50% formic acid.

Sample: Human hair or sheep wool.

Equipment: (i) Electric heating mantle (ii) Filtration unit (Büchner flask, sintered funnel, suction pump) (iii) Top loading balance (iv) Hot air oven (v) Rotary or flash evaporator.

Glassware: Reflux condenser along with 2 L round bottomed flask.

Procedure

(i)   Weigh 200 g of sheep wool or human hair and wash it repeatedly with chilled diethyl ether in a glass tray. (**Caution**: Keep ether away from open flame. Carry out this operation in a fume hood). Cut the dried specimen into small fragments with a sharp scissors and transfer it into a 2 L round bottomed flask. Add 500 mL of extraction reagent along with few porcelain chips and reflux the contents over a heating mantle for 18h.

(ii)   Transfer the hot contents immediately into a 1 L glass beaker and add 5 g of activated charcoal. Stir the contents vigorously with a glass rod and filter using a filtration unit. Collect the filtrate and concentrate it into thick syrup by flash evaporation. Dissolve the syrup in 100 mL of distilled water and add 50% hot sodium acetate solution, until the pH reaches 5.0 [the pH of the solution can be tested either by using a narrow range pH indicator paper or Congo red indicator (note the colour change from blue to red)].

(iii)   Store the contents at ambient temperature for 72h in order to precipitate cystine. Recover the material by filtration and wash it with warm distilled water. Dissolve the residue in 200-250 mL of warm 3% HCl. Mix the contents and filter. Collect the filtrate and add 2 g of activated charcoal. Stir and filter. If any colour persists, repeat the charcoal treatment.

(iv)   Cystine is precipitated by adding 50 % sodium acetate solution, until the *p*H reaches 5.0. Allow the contents to stand at ambient temperature for 6h. The precipitated amino acid is filtered and washed with warm water (50 mL x 5). Transfer the residue on to a glass Petri plate/ watch glass and dry overnight at 70°C in a hot air oven.

(v)   The dried material is dissolved in a minimum volume of ammonium hydroxide solution. Any undissolved matter is removed by filtration. Cystine present in the clear filtrate is precipitated by cautious addition of conc. HCl. Recover the crystalline precipitate by filtration followed by washing with ethanol and ether. Dry the recovered sample at ambient temperature in a Petri plate and weigh the content. Calculate the % yield.

Workout: (i) Perform qualitative tests for sulphur containing amino acids *(see 6. Qualitative Analysis)*. (ii) Check the purity of the preparation either by paper chromatography or TLC, using authentic standards of L-cystine and L-cysteine.

Experiment #3: Isolation of casein from milk by isoelectric precipitation

Principle: The solubility of most proteins, specially the globular proteins, is profoundly influenced by the *p*H of the aqueous medium. The *p*H at which a protein is least soluble is its isoelectric *p*H, wherein the net charge on the protein molecule is zero. Under these conditions there is no electrostatic repulsion between neighbouring protein molecules and they tend to coalesce and precipitate. However, it should be noted that, at *p*H values above or below the isoelectric point, all protein molecules have a net charge of the same sign. Thus, they repel each other, preventing coalescence of single molecules to insoluble aggregates.

Casein, the principle phosphoprotein of milk is separated from other proteins by isoelectric precipitation, i.e., by adjusting the pH of milk to its isoelectric pH (4.8).

Reagents: (i) Hydrochloric acid (0.5 N) (ii) Solvents: Ethanol, diethyl ether.

Sample: Cow or Buffalo milk.

Equipment: (i) pH meter (ii) Filtration unit (Büchner funnel, side arm flask, suction pump) (iii) Top loading balance (iv) Centrifuge.

Procedure

(i)   Centrifuge 100 mL of milk at 4000 rpm for 20 - 25 min at room temperature and carefully remove the cream/fat from the surface with the help of a stainless steel spatula.

(ii)   Transfer this skimmed milk into a 500 mL glass beaker. Add an equal volume of distilled water and stir. Note its pH using a pH meter.

(iii) Drop the pH of the diluted milk to 4.8 by drop wise addition of 0.5 N HCl using a burette, with constant stirring. At this pH, casein gets precipitated. Allow the precipitate to settle at room temperature for 30 min.

(iv)   Decant the supernatant carefully and filter the suspension using a filtration unit connected to a suction pump (Büchner funnel fitted with Whatman No.1 filter paper disc or through a sintered glass funnel).

(v)   The moist precipitate is washed thrice with 100 -150 mL of distilled water to remove the salts. This is followed by two washes each with 100 mL of ethanol and diethyl ether.

(vi)   Transfer the cake from the funnel onto a clean Petri-plate or a watch glass. Spread the material uniformly and allow it to dry at room temperature overnight.

(vii) Record the weight of the dried casein sample and calculate the percent yield.

Caution: Bottle containing diethyl ether should be chilled before opening. Keep away from flame.

Note: See appendix for isoelectric points of proteins.

Workout: Determine the nitrogen content of the isolated casein *(see 13. Food Analysis)*.

Experiment #4: Preparation of egg albumin

Principle: Albumin, the water soluble protein, is isolated from egg white by salting out technique. Globulins precipitate at half saturation, while albumins precipitate at full saturation with ammonium sulphate salt. Proteins in solution form hydrogen bonds with the surrounding aqueous *milieu*, through their exposed polar/ionic functional groups. When high concentration of small and highly charged ions such as ammonium and sulphate are added to protein solutions, they compete with protein for binding with water molecules. This interaction lowers the solubility of proteins, resulting in their precipitation. This forms the basis of the isolation of egg albumin.

Chemicals: (i) Solid ammonium sulphate.

Sample: Hen's egg.

Procedure

(i) Collect egg white from 2 eggs carefully, avoiding the egg yolk, into a 500 mL beaker. Dilute the egg white to 100 mL by adding distilled water with vigorous beating and stirring. Any precipitate (of ovomucoid) formed at this stage is removed by centrifugation. Add 32.5 g of solid ammonium sulphate in small proportions, at room temperature with gentle stirring (avoid frothing). Equilibrate the contents for 15 minutes at room temperature. The precipitate formed is that of globulins, which is removed by filtration or centrifugation.

(ii)   Saturate the filtrate or the supernatant by adding solid ammonium
       sulphate (35.5 g). Leave the contents at room temperature for 30
       minutes with intermittent stirring. The albumin precipitate formed is
       recovered by centrifugation. Dissolve the precipitate in minimum
       volume of distilled water and dialyse the protein solution extensively
       in cold, against distilled water in a 2 L plastic beaker (two to three
       changes of distilled water over a period 36 - 48h) so as to remove the
       salt (test the presence of ammonium sulphate in water by adding barium
       chloride solution (5% w/v) (a white precipitate of barium sulphate is
       observed). Measure the volume of the protein solution after dialysis
       and add 0.05% (w/v) sodium azide as preservative. Store the protein
       solution in cold.

Workout: Estimate the protein content in the sample by biuret method and
calculate the protein yield.

Experiment #5: Isolation of bovine liver catalase

Principle: Catalase, a hydro-peroxidase is a heme containing protein which
is widely distributed in nature. It decomposes hydrogen peroxide into water
and oxygen.

$$H_2O_2 \xrightarrow{\text{Catalase}} H_2O + 2O_2$$

The enzyme is isolated from the tissue by extraction with water, followed
by precipitation of the enzyme with organic solvent.

Reagents: (i) Chloroform : Ethanol reagent (1:2 v/v).

Sample: Bovine liver.

Procedure

(i)   Fresh bovine liver collected from a local abattoir in an ice bucket
       (4°C) is brought to the laboratory and washed with ice cold normal
       saline. Mince 500 g of ice cold liver in a mixer-grinder for about 1-2
       minutes. Transfer the liver mince into a 1L glass beaker. Extract the

enzyme by adding 500 mL of distilled water at room temperature by occasional stirring, for 10 minutes. Later, add 240 mL of chilled chloroform:ethanol reagent. Mix the contents vigorously for 30 seconds and filter through a funnel plugged with glass wool. Collect the filtrate and allow it to stand at room temperature, overnight.

(ii)   A sediment of catalase crystals is observed. The crystals are recovered by centrifugation (3500 rpm for 10 minutes). The crystalline sediment is dissolved in minimum volume of distilled water (10 -15 mL) and the solution is clarified by centrifugation to remove impurities (at 3500 rpm for 15-20 minutes). The decanted supernatant is chilled in a refrigerator overnight (16-24 hours). Catalase separates out as needle shaped crystals. Filter the contents and dry the material. Record its weight.

Workout: Assay the enzyme activity and determine its specific activity *(see, 10. Enzyme Assays)*.

Experiment #6: **Preparation of urease from jack bean meal**

Principle: The enzyme urease is distributed in plants and bacteria. It catalyses the hydrolysis of urea to ammonia and carbondioxide.

$$NH_2\text{-}CO\text{-}NH_2 \xrightarrow[H_2O]{Urease} 2NH_3 + CO_2$$

The enzyme is isolated from jack bean meal (*Canavalia ensiformis*) by aqueous acetone extraction and crystallization of the protein in cold.

Reagents: (i) Extraction solvent: Aqueous acetone (31.6% v/v).

Sample: Commercial Jack bean (*Canavalia ensiformis*) meal.

Procedure

(i)   Weigh and transfer 50 g of jack bean meal powder into a 500 mL glass beaker. Add 250 mL of the extraction solvent and stir the contents for 5 - 10 minutes and filter. Collect and store the filtrate in a refrigerator, overnight.

(ii) Next day, centrifuge the contents in a cold centrifuge (at 5000 rpm for 15 minutes). Decant the supernatant and wash the pellet with cold extraction solvent (5 -10 mL). Centrifuge again and discard the supernatant. Observe the shape of the crystals under a microscope.

Work out: (i) Assay the enzyme activity by dissolving the crystals in deionised water (20- 25 mL) (ii) Using the procedure, isolate urease from horse gram (*Dolichos biflorus*) and compare the enzyme activity *(see, 10. Enzyme Assays)*.

Experiment #7: Isolation of lactose from bovine milk

Principle: Lactose (milk sugar), the reducing disaccharide consisting of galactose and glucose as sugar units is linked by a β- 1- 4 glycosidic bond. The sugar is isolated from milk by acid and salt precipitation of milk proteins, followed by ethanol treatment.

Reagents: (i) Acetic acid (10% v/v) (ii) Ethanol (95% v/v) (iii) Ethanol (25% v/v) (iv) Chemicals: Calcium carbonate and activated charcoal

Sample: Cow or Buffalo skimmed milk.

Procedure

(i) Take 100 mL of milk sample in a 250 mL glass beaker and add 10%(v/v) acetic acid solution drop wise until there is significant precipitation of casein (avoid excess addition of acid, as this would promote the hydrolysis of lactose). Stir the contents and remove the precipitate by filtration or centrifugation (at 3500 rpm for 20 minutes). Collect the filtrate or the supernatant.

(ii) To the clear filtrate or the supernatant, add 2.5 – 3.0 g of solid calcium carbonate. Heat the contents by gentle boiling for 10 minutes over a hot plate. This step will lead to the precipitation of lactalbumin. Cool to room temperature and filter the solution to remove the precipitated albumin and undissolved salt.

(iii) Concentrate the filtrate to 10 to 15 mL by gentle boiling or by flash evaporation. To the hot concentrate, add 85 mL of 95% (v/v) ethanol and one gram of charcoal. Mix the contents thoroughly and filter. Collect the clear filtrate in a pear shaped flask. Allow the contents to stand for 3 to 5 days at room temperature, so as to allow the lactose to crystallize. Later, collect the lactose crystals by filtration. Wash the crystals with 5 - 10 mL of 25% (v/v) ethanol. Dry the sugar crystals and record their weight.

Workout: (i) Perform the reducing test with *Benedict's* reagent (ii) Prepare an osazone and observe the shape of the crystals under a microscope (iii) Check the purity of the isolated sugar by paper chromatography (iv) Estimate the sugar content by phenol-sulhpuric acid procedure *(see 7. **Quantitative Analysis)*** and calculate the % yield.

Experiment #8: Isolation of glycogen from liver

Principle: Glycogen, the branched chain homopolysaccharide is present both in liver and muscle. Glycogen can be effectively extracted from liver by treatment with alkali followed by separation of glycogen by alcoholic precipitation.

Reagents: (i) Potassium hydroxide (30%, w/v) (ii) Ethanol (95% v/v).

Sample: Rat or bovine liver.

Procedure

(i) Collect liver tissue from a well-fed albino rat (body mass = 250g) after treating it with ether as anaesthesia using a glass desiccator. Remove its liver by dissecting the rat on a dissection tray. Alternatively, collect bovine liver from a local abattoir, immediately after is slaughtered the animal is slaughtered. The tissue should be brought to the laboratory in an ice bucket.

(ii)   Clean the liver tissue with ice cold saline to remove blood. Any unwanted connective tissue should be removed by using a scissors. Weigh 10 g of liver tissue and cut it into small pieces using a scalpel or a razor. Transfer the minced liver into a 100 mL conical flask and add 20 mL of 30% KOH and mix. Heat the contents on a boiling water bath for 30 minutes with intermittent shaking and cool to room temperature.

(iii)   The contents are chilled in an ice bucket and precipitation of glycogen is effected by the addition of two volumes (40 mL) of ethanol (95%), with constant stirring. Allow the contents to stand in cold for 10 - 15 minutes in an ice bucket. Centrifuge at 4000 rpm for 30 minutes. Discard the supernatant and collect the precipitate. The precipitate is dissolved in minimal volume of distilled water and reprecipitated by the addition of two volumes of ethanol (95%). The glycogen precipitate is recovered by centrifugation at 4000 rpm for 15 minutes.

(iv)   Wash the above precipitate sequentially with chilled ethanol (15 mL) and diethyl ether (15 mL). Discard the supernatant after centrifugation. Dry the glycogen sample in a desiccator containing calcium chloride as a desiccant.

(v)   Weigh and calculate the % yield of glycogen.

Workout: Estimate the total sugar content by phenol-sulhpuric acid procedure *(see 7. **Quantitative Analysis**)*.

Experiment #9: Isolation of potato starch

Starch, a widely distributed plant reserve homopolysaccharide, containing α-glucopyranose as its repeating unit. It exists in two forms namely, amylose (unbranched) and amylopectin (branched), which vary in different amounts in starches. The amylopectin content in potato starch is about 70%.

Principle: Potato starch is isolated by extraction with water followed by filtration.

Sample: Fresh, cleaned potatoes.

Reagents: (i) Ethanol (95 %v/v).

Procedure

Weigh 100 g of peeled potatoes and cut them into small pieces with a knife. Transfer them into a homogenizer and add 250 mL of distilled water and homogenize. Transfer the homogenate into a 500 mL glass beaker. The pulverized material is then allowed to settle. Starch rapidly settles at the bottom. Decant the supernatant carefully and wash the residue, twice with distilled water and finally with ethanol. Filter through muslin cloth and air dry the sample. Express the recovery as g%.

Workout: (i) Perform the qualitative test for carbohydrates on the isolated sample and draw your conclusions. (ii) Isolate the starch from tapioca or sweet potato and compare the yields with that of potato.

Experiment #10: Isolation of cholesterol & lecithin from egg yolk

Principle: Egg yolk is a good source of cholesterol and phospholipids (ex. lecithin). These lipids are isolated from yolk by fractionation with organic solvents.

Sample: Hen's egg (one).

Reagents: (i) Acetone (ii) Petroleum ether (65 - 80°C boiling point).

Procedure

(i)   Take a boiled egg and collect the egg yolk (yellow coloured) carefully into 150 mL glass beaker. Add 40 mL of acetone and stir the contents with a glass rod. Allow the solids to settle. Decant the acetone extract in a 100 mL glass beaker. Repeat the acetone extraction once again. Pool the acetone extracts containing cholesterol into a glass container and store.

(ii)   Remove the acetone present in solid residual fractions by evaporation over a steam bath (under a fume hood). Cool the container and add 30 mL of cold petroleum ether to it, stir the contents vigorously and allow the solids to settle. Decant the petroleum ether extract containing lecithin in a glass container. Repeat the extraction procedure once again and pool the extracts.

(iii)   Individually evaporate both the extracts over a steam bath, using a chinadish in small volumes (in a fume hood). Crude cholesterol precipitates as light yellowish crystals, while lecithin separates as a yellowish-brown waxy liquid.

Workout: (i) Perform the qualitative tests for cholesterol (ii) Estimate the cholesterol content by *Liebermann-Burchard* reaction and calculate the yield (iii) Analyse the isolated lecithin along with reference standard by silica gel-TLC, using chloroform : methanol : water (65:25:4) as the developing solvent system. Identify the separated lipids by placing the developed TLC plate in a glass chamber containing iodine vapours. *(see 8. Biochemical Separation Techniques)*.

Caution: Both the solvents are highly inflammable. Keep the solvents away from flame.

Experiment #11: (a) Isolation of DNA from bovine spleen

Principle: DNA is present in all cell types. It is disintegrated into small fragments by the action of deoxyribonuclease (DNAse). An ideal source for isolation of mammalian tissue DNA is either thymus or spleen, as these tissues are rich in DNA and have low DNAse activity.

Reagents: (i) Sodium citrate (0.2 M, *p*H 7.4) -saline buffer, containing 0.9% (w/v) sodium chloride (ii) Sodium chloride solution 2.6 M (iii) ethanol 70% (v/v).

Sample: Bovine spleen - obtained from a local abattoir.

Procedure

(i)   Weigh 50 g of fresh and cleaned bovine spleen. Dice the tissue using a scalpel or a sharp blade. Transfer the diced tissue into a mechanical blender or a tissue homogeniser and add 200 mL of ice cold sodium citrate-saline buffer. Macerate the tissue thoroughly for 1-2 minutes in cold. Centrifuge the contents at 3500 rpm in cold for 20 minutes. The pellet obtained contains most of the DNA, cell debris and unbroken cells, while the supernatant has soluble proteins, carbohydrates and RNA. Supernatant is discarded and the pellet is processed further.

(ii) Suspend the pellet in 200 mL of sodium citrate buffer, 0.2 M, *p*H 7.4, and homogenise, once again. Collect the pellet by centrifugation and discard the supernatant. Resuspend the pellet in 60 mL of ice-cold sodium chloride (2.6 M) solution by breaking the pellet with a glass rod followed by vortexing the content vigorously. This process results in dissociation of nucleic acid from the nucleoprotein, followed by its solubilization. Centrifuge the contents at 5000 rpm for 10 minutes, in cold and collect the supernatant. Resuspend the pellet in 60 mL of ice-cold sodium chloride (2.6 M) solution and repeat the extraction procedure. Pool both the supernatants and recentrifuge the contents at 20,000 x g for 20 minutes in cold. (iii) Decant the supernatant and measure its volume. Transfer the supernatant carefully into a 500 mL beaker, kept in an ice bucket. Add twice the volume of ice cold ethanol, allowing it to pour down the side of the beaker. Collect the fibrous DNA by carefully inserting a bent glass rod (J - shaped) all the way to the bottom and stirring in small circles in a clockwise direction. The spooled DNA is removed from the beaker and washed twice with 25 - 30 mL of cold ethanol (70% v/v). The DNA sample is dissolved in 5 mL of sodium citrate-saline buffer and stored in cold.

Workout: (i) Check the purity of the DNA preparation by recording the UV absorbance at 260 nm and 280 nm. Calculate the ratio of absorbance at 260 nm / 280 nm (ii) Estimate the DNA content by diphenylamine procedure, using calf-thymus DNA as reference standard *(see 7. **Quantitative Analysis**)*.

Note: The purity of the DNA preparation can be improved by enzymatic treatment with pronase (a proteolytic enzyme), amylase and RNAse.

Experiment 11: (b) **Isolation of DNA from onion**

Sample: Fresh onions (*Allium cepa*).

Reagents: (i) Extraction buffer: ethylenediaminetetra acetic acid-sodium salt (EDTA), 0.1 M, containing 0.9%(w/v) sodium chloride solution, pH 7.0 (ii) sodium dodecyl sulphate (SDS) 20% (w/v) solution (iii) sodium perchlorate ($NaClO_4$) 5 M solution (iv) chloroform : isoamyl alcohol (24:1) solvent (v) ethanol (95% v/v) (vi) ethanol (70% v/v) (vii) sodium citrate (0.2 M, pH 7.4) -saline buffer containing 0.9% (w/v) sodium chloride.

Procedure

(i)   Weigh 50 g of fresh peeled and clean onion tissue. Cut the tissue into small pieces with a knife and transfer the same into a mechanical blender and homogenise by adding 150 mL of ice-cold extraction buffer for 3-5 minutes. Extraction of DNA in the presence of EDTA, chelates metal ions, which in turn protect DNA from degradation.

(ii)   Pour the contents into a 500 mL glass beaker. Add 15 mL of SDS solution. Mix and heat the contents for 10 minutes at 60°C in a hot water bath. Cool the contents to room temperature and centrifuge at 3000 rpm for 15 minutes. Use of SDS aids in disrupting the cell membranes and brings about denaturation of proteins. Measure the volume of the supernatant and add 5 M perchlorate solution, so that the final concentration of perchlorate is 1M. This step disrupts the DNA-protein interaction.

(iii) Transfer the perchlorate treated solution into a stoppered conical flask and add an equal volume of chloroform:isoamyl alcohol solvent. Gently shake the contents for 30 minutes using a wrist arm shaker. This step results in the formation of three phases: (a) an upper aqueous phase containing DNA, RNA and carbohydrates (b) interphase containing denatured proteins and (c) lower organic phase containing lipids.

(iv)   Carefully separate the upper aqueous phase and transfer it into a glass beaker kept in as ice bucket. Add twice the volume of ice cold ethanol, allowing it to pour down the side of the beaker. Collect the fibrous DNA by carefully inserting a bent glass rod (J - shaped) all the way to the bottom and stirring in small clockwise circles. The spooled DNA is removed from the beaker and washed twice with 25

- 30 mL of cold ethanol (70%). The DNA sample is dissolved in 5 mL of sodium citrate-saline buffer and stored in cold.

Workout: (i) Estimate the amount of DNA by diphenylamine method and calculate the % yield. (ii) Determine the purity by spectral analysis.

Experiment #12: Isolation of yeast RNA

Principle: Ribonucleic acid (RNA) is distributed both in cytoplasm and nucleus of the yeast cell. RNA is isolated from the whole cell homogenate by phenol extraction. Phenol treatment leads to disruption of protein-nucleic acid interaction and denaturation of proteins. Later RNA is isolated by ethanol precipitation.

Reagents:(i) Freshly distilled phenol (ii) Ethanol (iii) Ethanol: diethyl ether (3:1) mixture (iv) Diethyl ether (v) Potassium acetate solution (20% w/v), *p*H adjusted to 5.2.

Caution: Phenol : Corrosive chemical

Sample: Commercial Baker's Yeast.

Procedure

(i)   Weigh 50 g of the yeast powder and transfer it into a 1 L glass beaker and add 200 mL of distilled water. Leave the contents aside for 15 minutes at 37°C, later add 267 mL of distilled phenol and stir the contents mechanically for 30 minutes at room temperature. Centrifuge the contents in cold at 3500 rpm for 15 minutes. After centrifugation, three phases are observed, (a) lower phenolic phase consisting of DNA, (b) upper aqueous phase containing RNA and (c) interphase having denatured proteins.

(ii)   The aqueous phase is carefully collected and recentrifuged at 7000 x g for 20 minutes, in cold. Residual phenol in the aqueous phase is removed by washing it with 50 mL of diethyl ether, twice. Decant the supernatant and measure the volume. For every 90 mL of the supernatant, mix 10 mL of 20% (w/v) potassium acetate solution (final concentration of potassium acetate is 2%). Precipitate, the RNA by adding two volumes of cold ethanol. Leave the contents, overnight in a refrigerator (0-4°C). Next day, centrifuge the contents in cold for 15 minutes at 3500 rpm and collect the pellet. Wash the pellet with ethanol:ether mixture, followed by ethanol and finally with ether, centrifuging after each step. Dry the sample at room temperature and record its weight.

Workout: (i) Estimate the RNA content of the preparation by orcinol method and calculate the yield *(see, 7. Quantitative Analysis)* (ii) Determine the UV-absorption spectra of the sample in sodium citrate-saline buffer (100 mM, *p*H 7.4).

Caution: Phenol is corrosive, should be handled with gloves. Use, pro-pipette for dispensing.

Additional Reading

1.   Carter, H. E. (ed). Biochemical Preparations. Vol. 1. John Wiley & Sons. Inc. NY, USA/ Chapman & Hall, London, (1950).

2.   Colowick, S. P. & Kaplan, O. N. (eds). Methods in Enzymology. Vol. II. Academic Press Inc. Publishers, NY, USA, (1955).

3.   Vestling, C. S. (ed). Biochemical Preparations. Vol. 6. John Wiley & Sons. Inc. NY, USA/ Chapman & Hall, London, (1958).

4.   Greenstein, J. P. & Winitz, M. Chemistry of the Amino acids. Vol. 3. John Wiley & Sons. Inc. (1961).

5.   Anfinsen, C. B. Jr., Anson, M. L., Edsall, J. T. & Richards, F. M. (eds). Advances in Protein Chemistry. Vol. 20. Academic Press, NY, USA, (1965).

6.   Hawk, P.B. Oser, B L. & Summerson, W. H. Practical Physiological Chemistry. 14th ed. Churchill, London, (1966).

7.   Anfinsen, C. B. Jr., Anson, M. L., Edsall, J. T. & Richards, F. M. (eds). Advances in Protein Chemistry. Vol. 22. Academic Press, NY, USA, (1967).

8.   Christie, W. W. Lipid Analysis. Pergamon Press, NY, USA, (1982).

9.   Dashek, W. V. (ed). Methods in Plant Biochemistry and Molecular Biology. CRC Press, Boca Raton, USA, (1997).

10.   Dickson, C. Medicinal Chemistry Laboratory Manual. CRC Press, Boca Raton, USA, (1999).

Lab Notes

Lab Notes

Enzyme Assays

Enzymes are the biocatalysts produced by living cells and catalyze numerous metabolic reactions occurring within the cell. They are proteinaceous in nature (exception - *ribozyme*). Enzymes influence the rate or the speed at which a biochemical reaction attains equilibrium. They participate in the biological reactions by providing a reaction path that lowers the activation energy for the transition of substrate to products at relatively mild environmental conditions.

In contrast to non-enzymatic catalysts enzymes exhibit a high degree of substrate specificity. Only a small region of the enzyme molecule, referred to as the *active site* of an enzyme, is involved in the binding of the substrate and its subsequent catalysis. The three-dimensional structure of the active site is complementary to that of the spacial arrangement of the substrate molecule. Various factors such as pH, temperature, ionic strength, enzyme substrate and product concentration influence the enzyme activity. Following equation gives the simple relationship of an enzyme catalysed reaction:

$$E + S \longleftrightarrow ES \longleftrightarrow EP \longleftrightarrow E + P$$

Where, E = Enzyme; S = Substrate; ES = Enzyme substrate transient complex; EP = Enzyme product transient complex; P = Product.

ENZYME CLASSIFICATION

As per the recommendations of the Enzyme Commission, (IUB), enzymes are divided into six major groups on the basis of the type of the reaction they catalyse.

Enzyme Class	General reaction
1. *Oxidoreductases*: Catalyse oxidation-reduction reactions. Ex: Lactate Dehydrogenase.	$AH_2 + X \Rightarrow A + XH_2$
2. *Transferases*: Catalyse transfer of groups (amino, methyl, phosphoryl etc.,) Ex. Transaminase.	$A\text{-}X + B \Rightarrow B\text{-}X + A$
3. *Hydrolases*: Catalyse hydrolytic cleavage of covalent bonds. Ex. Peptidases (cleave peptide bonds).	$A\text{-}B + H_2O \Rightarrow AH + BOH$
4. *Lyases*: Catalyse cleavage of bonds without addition of water. Ex. Aldolase.	$A\text{-}B \Rightarrow A + B$
5. *Isomerases*: Catalyse intramolecular rearrangement. Ex. Phosphohexoisomerase.	$X \Rightarrow X^*$
6. *Ligases*: Catalyse the synthesis of a bond coupled to ATP hydrolysis. Ex. Glutamine synthase.	$A + B \Rightarrow A\text{-}B$ ATP ADP+Pi

UNITS OF ENZYME ACTIVITY

Normally, in most of the biological preparations the actual concentration of the enzyme is difficult to determine, hence the amount of enzyme present can be expressed in terms of its activity.

One international unit (IU): An enzyme unit (IU) is that concentration of an enzyme that catalyses the formation of 1 μmole of product per minute under defined assay conditions. The concentration of enzyme in crude biological preparation is expressed as units/mL (U mL^{-1}).

For uniformity in reporting the enzyme activities, the Enzyme Commission of IUB, has recommended a new unit namely, *Katal*. One katal refers to the conversion of one mole of substrate per second. The enzyme activity can be expressed as millikatals, microkatals and nanokatals. Thus, one IU of enzyme = 1/60 μkatal = 16.67 nkatal.

Specific activity is expressed as the number of enzyme units per milligram protein. It is an index of enzyme purity. During the process of enzyme purification, the value of specific activity increases. A pure enzyme preparation exhibits maximal specific activity.

Total activity of an enzyme is calculated by (Specific activity) x [total protein (mg) in the enzyme preparation].

MEASUREMENT OF ENZYME ACTIVITY

Usually, biochemical analysis involves determination of the concentration of an analyte in biological specimen. Unlike these quantitative measurements, the measure of an enzyme molecule is based on their biological activity rather than the measurement of their actual concentration. In determining the enzyme activity, the relationship between the product formed from the substrate and the absorbance due to increase in concentration of the product is a quantitative measure of the velocity or the rate of the reaction as a function of time. Further, the concentration of the product formed can be calculated by using the physical factor, the molar extinction coefficient (ε) (as detailed in *4. Basic concepts and use of instruments in biochemical analysis*). Enzyme activity can be assayed by two methods, namely (i) equilibrium or steady- state and (ii) kinetic methods. Figure 10.1 shows a typical enzymatic reaction. In the former, the enzyme is incubated with its substrate for a pre-determined time, at the end of which the enzymatic reaction is stopped and the amount of product formed or substrate consumed is measured. In this type of assay, it is necessary to establish the linear response range, over which the zero order kinetics is followed. If this aspect of the assay is disregarded, the measurements made would lead to an erroneous interpretation.

There are two approaches to assay an enzyme by steady-state method *viz.,* one-point and two-point assays. One-point assay is based on a single pre-determined time, while the two-point method has two pre-determined time points, representing the linear response of the enzymatic reaction. This is illustrated in figure 10.2. In kinetic assay, the reaction rate is continuously monitored at short intervals of time. Hence, any deviation from zero order kinetics can be noted by a non-linear rate of reaction. Kinetic assays are potentially more accurate than steady-state methods of enzymatic analysis.

Experiment #1 : **Acid phosphatase** (EC.3.1.3.2, *o- phosphoric monoester phosphohydrolase, p*H optima acid) **assay.**

Principle: The enzyme acid phosphatase catalyses the conversion of the

Figure 10.1 A typical enzymatic reaction.

Figure 10.2 A typical two-point enzyme
assay based on absorbance-time relationship.

substrate, p-nitrophenyl phosphate to p-nitrophenol (a yellow coloured product) and inorganic phosphate. The enzyme activity is measured by recording the absorbance due to the formation of p-nitrophenol, at 410 nm. The concentration of p-nitrophenol is calculated by using its molar extinction coefficient value at λ_{max} of 410nm.

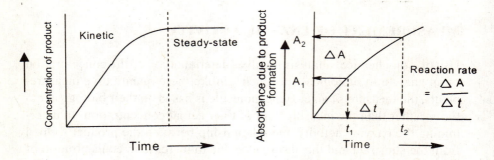

p-nitrophenyl phosphate p-nitrophenol   Inorganic phosphate

Enzyme unit: One enzyme unit hydrolyses 1 µmole of p-nitrophenyl phosphate per minute, at pH 4.8 at 37°C.

Reagents: (i) Sodium citrate buffer, 100 mM, pH 4.8. (ii) Substrate solution: p-nitrophenyl phosphate, 15 mM dissolved in deionised or double glass distilled water (iii) Sodium hydroxide (0.2 N).

Enzyme source: Fresh potato (*Solanum tuberosum*).

Procedure

(i) Enzyme extraction: Take 50 g of fresh washed potato and peel the skin. Cut it into small pieces with a sharp stainless steel knife. Weigh 20 g of the cut potatoes and transfer it into a mechanical blender along with 80 mL of cold deionised or double glass distilled water. Homogenize the contents thoroughly for 3 - 5 minutes. Centrifuge the homogenate in cold at 4000 rpm, for 15 minutes to obtain a clear supernatant. Decant the supernatant into a clean 100 mL conical flask and store in cold. Use this fraction as a source of enzyme.

(ii) Enzyme assay: Pipette out the following reagents into clean labelled test tubes. Run a blank along with the test sample, in triplicates.

Enzyme assay protocol

Reagents	Blank (mL)	Test (mL)
1. Citrate buffer, 100 mM, pH 4.8	0.5	0.5
2. Substrate solution (p-nitrophenyl phosphate, 15 mM)	0.5	0.5
Mix the contents thoroughly by vortexing.		
3. Enzyme extract	-	0.1
Mix and incubate at 37°C in a water bath exactly for 10 minutes.		
4. Sodium hydroxide solution (0.2 N)	3.9	3.9
Mix the contents		
5. Enzyme extract	0.1	-
Total assay volume	5.0	5.0

Record the absorbance of test and blank at 410 nm, using a spectrophotometer.

(iii) Calculation: Using the following equation, calculate the activity, i.e., U mL^{-1} of the enzyme solution,

$$\text{Units/mL enzyme extract} = \frac{(\text{Absorbance of test} - \text{Absorbance of Blank}) \times 5.0}{10 \times 18.3 \times 0.1} \times 10^*$$

Where,

5.0 = Total enzyme assay volume in mL.

10   = Time of assay in minutes.

18.3 = Millimolar extinction coefficient (ε) of *p*-nitrophenol at 410 nm.

0.1 = Volume of enzyme extract.

*   = Factor

Note: (i) If the colour developed is too intense, dilute the enzyme extract appropriately and repeat the enzyme assay. Take the dilution factor into account while calculating the enzyme activity (ii) If the colour development is insufficient for 10 minutes, increase the incubation time to 20 minutes.

Workout: Calculate the specific activity of the enzyme by the following equation:

$$\text{Specific activity} = \frac{\text{Units/mL enzyme extract}}{\text{mg protein/ mL enzyme extract}}$$

Note: Determine the protein content in the enzyme extract by Lowry method, for calculating the specific activity.

Experiment #2: **Assay of β-amylase** (EC 3.2.1.2, α- *1,4- glucan maltohydrolase).**

Principle: The enzyme β-amylase catalyses the hydrolysis of α-1 → 4, glycosidic linkages from the non-reducing end of polysaccharides (starch - amylose, amylopectin and glycogen), to yield maltose units. This enzyme does not hydrolyse α-1 → 6, glycosidic linkages present in branch chain polysaccharides (amylopectin, glycogen). The activity of the enzyme is measured colorimetrically by estimating the amount of reducing sugar formed (maltose), by 3, 5-dinitrosalicylic acid (DNS) method.

Enzyme unit: One unit of enzyme will liberate 1.0 mg of maltose from 1% starch in 5 minutes under defined conditions.

Reagents: (i) Sodium acetate buffer 100 mM, pH 4.8, (prepare in deionised water) (ii) Sodium phosphate buffered saline, 20 mM, pH 7.0, containing 0.9 % sodium chloride, (iii) Soluble starch solution: Starch 1 % (w/v) in deionised water, (iv) DNS reagent: It is prepared by dissolving 1 g of DNS, 200 mg of phenol and 50 mg sodium sulphite in 100 mL of 1% (w/v) sodium hydroxide solution (prepare in deionised water), (v) Potassium sodium tartrate solution: Dissolve 30 g (w/v) of potassium sodium tartrate and make up the volume in 100 mL of 0.25 M sodium hydroxide solution.

Enzyme source: Sweet potato (*Ipomoea batates*).

Procedure:

(i) Enzyme extraction: Take 10 g of fresh, cleaned and peeled sweet potato. Cut it into small pieces using a stainless steel knife. Homogenize the tissue in 40 mL of cold sodium phosphate buffer, 20 mM, pH 7.0, containing 0.9% sodium chloride, using a mechanical homogenizer for 3 - 5 minutes. Centrifuge the homogenate in cold at 10000 to 12000 x g for 20 minutes. Collect the supernatant and use as an enzyme source. Store the enzyme extract at 4°C until use. Run a blank along with the test sample, in triplicates.

(ii) Pipette the following into clean labelled test tubes:

Reagents	Blank (mL)	Test (mL)
1. Sodium acetate buffer, 100 mM, pH 4.8	1.0	0.8
2. Starch solution (1%)	1.0	1.0
Mix the contents thoroughly by vortexing.		
3. Enzyme extract	-	0.2
Mix and incubate at 30°C in a water bath exactly for 10 minutes.		
4. DNS reagent	2.0	2.0
Mix and heat the tubes in a boiling water bath for 5 minutes.		
5. Potassium sodium tartrate reagent (add while the contents are still warm)	1.0	1.0
6. Distilled water	7.0	7.0
Total assay volume	12.0	12.0

(iii)   Record the absorbance of test and blank at 540 nm. Construct a
        standard curve for maltose (standard range 0.2 to 2 mg maltose) using
        DNS method (*see 7 . Quantitative analysis*). Calculate the amount
        of maltose liberated by the enzymatic reaction from the calibration
        plot.

$$\text{Calculation: Units/mL enzyme extract} = \frac{\text{mg of maltose released}}{\text{Volume of enzyme used (mL)}} \times 5$$

Workout: Estimate the protein content in the enzyme extract by Lowry
method and calculate the specific activity of the enzyme preparation.

$$\text{Specific activity} = \text{Units/mg protein} = \frac{\text{Units / mL enzyme extract}}{\text{mg protein / mL enzyme extract}}$$

Experiment #3: **Demonstration of catalase** (EC 1.11.1.6, *Hydrogen
peroxide: Hydrogen peroxide oxidoreductase*) **activity**.

Principle: The enzyme catalase catalyzes the decomposition of hydrogen
peroxide to water and oxygen.

$$2H_2O_2 \xrightarrow[pH\ 7.0]{\text{Catalase}} 2H_2O + O_2 \uparrow$$

This enzymatic reaction is monitored titrimetrically by the rate of
disappearance of hydrogen peroxide, using potassium permanganate, as
represented in the following equation:

$$2\,KMnO_4 + 5\,H_2O_2 + 3\,H_2SO_4 \longrightarrow K_2SO_4 + 2\,MnSO_4 + 8\,H_2O + 5\,O_2$$

The volume of potassium permanganate consumed in the titration is directly
proportional to the concentration of remaining (i.e., undecomposed)
hydrogen peroxide (**Note**: Two molecules of $KMnO_4$ react with five
molecules of H_2O_2).

Reagents: (i) Phosphate buffer, 50 mM, pH 7.0 (K^+/Na^+ (KH_2PO_4/ Na_2HPO_4) (ii) Substrate: Hydrogen peroxide (1.5%, 444 mM) (iii) Potassium permanganate solution (2% w/v, 126.6 mM) (iv) Sulphuric acid solution (1 M) containing 0.015 mM manganese sulphate as catalyst.

Equipment: All glass homogenizer, refrigerated centrifuge, pH meter, stop watch.

Enzyme source: Bovine/Porcine liver tissue.

Preparation of enzyme extract: Collect fresh bovine liver from local *abattoir* and transport to the laboratory in an ice bucket. Clean the tissue with ice cold phosphate buffer (0- 4°C). Weigh 1 g of wet liver tissue and transfer it into a clean all glass hand homogenizer (50 mL capacity). Add 25 mL of ice cold phosphate buffer and finely homogenize the tissue in cold condition, using an ice bath. Transfer the homogenate into a polypropylene centrifuge tube. Wash the glass homogenizer with 10 - 20 mL of cold buffer and pool it with the initial homogenate. Centrifuge the contents in a refrigerated centrifuge at 3000 x g, for 20 min. Collect and transfer the supernatant into measuring cylinder and make up the volume to 100 mL with cold phosphate buffer. Store the enzyme extract in an ice bucket or in a refrigerator. Prepare the enzyme extract freshly before the experiment for good results.

Procedure

(i) *Blank or base line titration*: Take 10 mL of hydrogen peroxide (1.5 %) into a 50 mL glass beaker and add 1 mL of distilled water and 10 mL of sulphuric acid (1.0 M) solution. Mix and transfer 5 mL of the sample into an Erlenmeyer flask and titrate it against $KMnO_4$ solution (2%) to a pale pink end point. Record the titer value. Repeat the titration thrice to get a consistent value. Calculate the initial concentration of hydrogen peroxide at zero time, by using the following equation:

$$\frac{V_1 \times M_1}{n_1} = \frac{V_2 \times M_2}{n_2} \qquad Equation....1$$

Where, M_1 = Molarity of H_2O_2 (to be calculated);

M_2 =   Molarity of $KMnO_4$ (i.e., 126.6 mM)

V_1 =   Volume of H_2O_2 used for titration

V_2 =   Volume of $KMnO_4$ consumed

n_1 =   Number of molecules of H_2O_2

n_2 =   Number of molecules of $KMnO_4$

Note: As per the above given chemical equation, two molecules of $KMnO_4$ reacts with five molecules of H_2O_2.

(ii)   Separately, perform blank titration for each time point (i.e., at 10, 30, 60, 120, 180 and 360 sec) as given in the Table 10.1. Calculate the millimolar (mM) concentration of H_2O_2 at each time point, by using equation-1. This accounts for non-enzymatic decomposition of H_2O_2, if any.

(iii) *Enzyme assay*: Take 10 mL of hydrogen peroxide (1.5 %) into a 50 mL glass beaker and add 1 mL of enzyme extract and swirl the beaker gently for 10 sec (note the time using a stop watch). Perform the enzyme assay at ambient temperature. At the end of 10 sec, stop the enzyme reaction by adding 10 mL of sulphuric acid (1.0 M) solution. Mix and transfer 5 mL of the sample into an Erlenmeyer flask and titrate the contents against $KMnO_4$ solution (2%) to a pale pink end point. Record the titer value (final titer value – initial titer = volume of $KMnO_4$ solution consumed, which is directly proportional to the remaining H_2O_2 in the assay system).

Repeat the above assay procedure and allow the enzymatic reaction to proceed for 30, 60, 90, 120, 180 and 360 sec., separately. At the end of each time point, determine the titer value as given in the table below.

(iv)   Calculate the remaining concentration of H_2O_2 at each time point, after the enzyme catalysed decomposition of hydrogen peroxide, using *equation -1*. The difference between the blank/baseline value at respective time point and the enzyme catalysed reaction gives the rate of decomposition of hydrogen peroxide as a function of time.

(v)   Draw a graph by plotting time (sec) on *x*- axis and concentration of

Table 10.1 Blank/Baseline titration at different time points

	Time (sec)						
	10	30	60	90	120	180	360
Blank (Baseline)							
1. Initial burette reading							
2. Final burette reading							
KMnO$_4$ consumed (mL)*							
Conc. of remaining H$_2$O$_2$ (mM)							

*The difference between the final and initial burette readings indicates the remaining hydrogen peroxide in the reaction mixture, after stopping the enzyme catalysed reaction at the end of respective time points. The shorter the incubation time of reaction, greater is the amount of KMnO$_4$ consumed during the titration.

decomposed H$_2$O$_2$ (in mM) on *y*- axis (figure 10.3).The rate of the enzymatic reaction is the slope of the linear portion of the curve. Calculate the rate of the reaction by taking two points on the linear portion of the graph and substituting the concentration of decomposed H$_2$O$_2$ (y_2 - y_1) and respective time points (t_2 - t_1) in the following equation:

$$\text{Reaction rate} = \text{Slope} = \frac{y_2 - y_1}{t_2 - t_1} = \frac{\Delta y}{\Delta t}$$

(The rate of the reaction can be expressed as µmoles of H$_2$O$_2$ decomposed/ sec under the specified experimental conditions).

Workout: From the experimental data find out the time point where the reaction rate is (i) lowest and (ii) highest.

Record the rates in the table below:

Time (sec) ⟶	0 -10	10-30	30-60	60-90	90-120	120-240	240-360
Rate of H$_2$O$_2$ decomposition/sec							

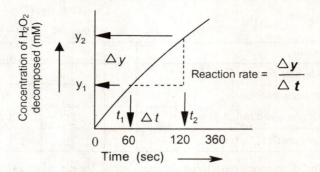

Fig. 10.3 Measuring the rate of disappearance of substrate (H_2O_2).

Experiment #4: Urease enzyme (EC 3.5.1.5, *Urea amidohydrolase*) **assay**.

Principle : Urea is split to ammonia and CO_2 by the action of urease and the liberated ammonia reacts with phenol in the presence of hypochlorite to form a blue coloured indophenol, in alkaline condition (*Berthelot reaction*). In this reaction, sodium nitroprusside acts as a catalyst. The blue colour developed is photometrically measured at 630 nm.

$$H_2N - \overset{\overset{\displaystyle O}{\|}}{C} - NH_2 \xrightarrow{\text{Urease}} CO_2 + 2NH_3$$
Urea

Enzyme unit: One enzyme unit liberates 1 µmole of ammonia from urea per minute under defined conditions of the assay.

Reagents : (i) Phosphate buffer (100 mM PB) , *p*H 6.8 (prepare the buffer in deionised water) (ii) Phenol Reagent : 50 g of phenol is dissolved in distilled water and made up to 1 L (5%). Store the reagent in an amber coloured reagent bottle. Prior to use, 0.25 g of sodium nitroprusside is added to phenol reagent (iii) Hypochlorite reagent (alkaline) : This acts as an oxidizing agent. Sodium hydroxide (25 g) is dissolved in 200 - 250 mL of distilled water, taken in a beaker (allow it to cool in an ice bucket), followed by addition of 40 mL of 5% sodium hypochlorite solution and the final volume is made up to 1 L with distilled water. Dilute 1 mL of alkaline hypochlorite reagent to 4 mL, before use (iv) Urea (1%) (freshly prepared in deionised water) (v) 7 N H_2SO_4.

Enzyme source: Jack bean (*Canavalia ensiformis*) or Jack bean meal: To 1.0 g of the powdered sample, add 25 mL of 100 mM phosphate buffer, *p*H 6.8 and gently mix to avoid frothing for 30 min. Centrifuge the contents at 3000 rpm, for 15 min. Decant the supernatant into a clean glass beaker and store in cold until further use. Take appropriate aliquots for the enzyme assay.

Standard ammonia solution (Stock -1.0 mg / mL) : Weigh 367.02 mg of ammonium sulphate and transfer into a 100 mL volumetric flask. Dissolve and make up to 100 mL with distilled water. Before use, dilute 10 ml of this solution to 100 mL with distilled water in order to obtain an ammonia concentration of 10 µg / mL.

Standard concentration range for analysis 1 - 10 µg/mL.

Procedure

(i) Protocol for enzyme assay: Dispense the following solutions, into clean and labelled test tubes,

Enzyme assay protocol

Reagents	Blank (mL)	Test (mL)
1. Phosphate buffer, 100 mM, *p*H 6.8	1.0	1.0
2. Urea solution (1%)	1.0	1.0
Mix the contents thoroughly by vortexing.		
3. Enzyme extract	-	0.5
Mix and incubate at 37°C in a water bath exactly for 30 minutes, with constant gentle shaking.		
4. Sulphuric acid (7 N)	1.0	1.0
5. Enzyme extract	0.5	-
Total assay volume	3.5	3.5

(ii) Procedure for ammonia estimation: To 1.0 mL of the standard solution containing 1 - 10 μg of ammonia, add 1.0 mL of phenol reagent, followed by1.0 mL of alkaline hypochlorite solution. Incubate the contents at 50°C for 5 min in a constant temperature water bath. The colour is read at 630 nm in a photometer.

(iii) Estimation of ammonia liberated during the enzymatic reaction: After completion of the enzymatic reaction (both blank and test), centrifuge the contents at room temperature for 15 min, at 3500 rpm. To 1 mL of the supernatant, add 4 mL of distilled water and mix. Take appropriate aliquots for ammonia estimation.

Protocol:

S.No.	Standard Ammonium sulphate (mL)	Distilled Water (mL)	Conc. of Ammonia (μg)	Phenol reagent (mL)	Alkaline-hypochlorite reagent (mL)	Absorbance at 630 nm
1.	Blank	1.0	- -	1.0	1.0	
2.	0.1	0.9	1.0	"	"	
3.	0.2	0.8	2.0	"	"	
4.	0.4	0.6	4.0	"	"	
5.	0.6	0.4	6.0	"	"	
6.	0.8	0.2	8.0	"	"	
7.	1.0	- -	10.0	"	"	
8.	Enzyme blank			"	"	
9.	Test			"	"	

" = same volume

(iv) Construct a calibration curve for ammonia (standard range 1.0 to 10 μg), by plotting the ammonia concentration (in μ moles) on x - axis and absorbance value at 630 nm on y- axis. Calculate the ammonia liberated by the enzymatic reaction from the calibration plot, after deducting the blank value from the test.

(v) Compute the urease activity and express the units/mL of the enzyme extract.

(**Note**: One enzyme unit liberates 1 μmole of ammonia from urea per minute at *p*H 6.8 at 37°C).

Workout: Determine the (i) optimal pH and temperature (ii) specific activity and the total activity of urease enzyme / g of horse gram (*Dolichos biflorus*) seed powder. Note: Use 5 % (w/v) seed extract, as a source of enzyme. For good results, use fresh seeds.

Procedure: *Affect of pH*: To 0.5 mL of the enzyme extract, add 1 mL of phosphate buffer 100 mM of varying *p*H (6.0, 6.5, 7.0, 8.0) and 1.0 mL of freshly prepared urea (1%) solution in different test tubes. Mix and incubate at various *p*H at 37°C for 30 min. Stop the reaction by adding 1mL of 7 N sulphuric acid. Centrifuge and use the supernatant for ammonia estimation. Dilute the supernatant before ammonia determination. Plot the enzyme activity *vs p*H and determine the *p*H optima.

Affect of temperature: To 0.5 mL of the enzyme extract, add 1 mL of phosphate buffer 100 mM, *p*H (use optimal *p*H, determined in the above experiment) and 1.0 mL of freshly prepared urea (1%) solution in different test tubes. Mix and incubate at various temperatures (25, 37, 45, 60 and 70°C) for 30 min. Stop the reaction by adding 1 mL of 7 N sulphuric acid. Centrifuge and use the supernatant for ammonia estimation. Dilute the supernatant before ammonia determination. Plot the enzyme activity *vs* temperature and determine the optimal temperature of the enzyme.

Additional Reading

1.   Thomas, E. B. Enzyme Hand Book, Vol. I & II. Springer-Verlag, Berlin, Germany, (1969).

2.   Segel, I. H. Biochemical Calculations. 2nd ed. John Wiley & Sons. Inc. New York. (1976).

3.   Bergmeyer, H. U. (ed-in-chief). Methods of Enzymatic Analysis. Vol. III & IV. 3rd ed. Verlag Chemie, Weinheim, Germany, (1983).

4.   Eisenthal, R. & Danson, N. J. (eds). Enzyme Assays: *A Practical Approach*. IRL, Press, Oxford, UK, (1992).

5.   Engel, P. C. Enzymology LABFAX. Academic Press Inc. San Diego, CA, USA, (1996).

Lab Notes

11

Clinical Biochemistry

In any diseased condition, pathological changes are preceded by biochemical changes. Thus, health of an individual depends on the harmonious equilibrium of all the biochemical reactions occurring in the body, while the diseased condition reflects the abnormalities in the normal biochemical reactions. Advances in the areas of metabolism, enzymology, pathology, medicine and analytical biochemistry have illuminated the development of clinical biochemistry as an applied science.

For a physician to design an appropriate therapy, biochemical tests are one of the valuable tools in discerning the cause of a disease, in addition to the patient's case history and clinical symptoms. The results of biochemical analysis are useful in (i) the assessment of diagnostic and prognostic indicators (ii) risk management (iii) monitoring the effectiveness of the treatment during the course of the therapy (iv) evaluation of drug toxicity and (v) screening of metabolic disorders of genetic origin, especially in neonates.

Biochemical investigations are carried out in a variety of biological samples. These include (i) whole blood (ii) plasma (iii) serum (iv) cerbrospinal fluid (v) serous fluid (ex. pleural, peritoneal fluids) (vi) mucosal secretions (ex. gastric juice) (vii) sweat (viii) saliva and (ix) urine. Table 11.1 gives the type of body fluid and the method of specimen collection for biochemical analysis.

A. BLOOD ANALYSIS

(i) *Preparation of blood serum & plasma*: Serum is prepared by centrifugation (at 2500 x g for 10 min) of whole blood which has been clotted in the absence of an anticoagulant. Serum separates as a straw coloured fluid. Clean, screw capped glass tubes are best suited for clotting of blood and its subsequent separation as serum. Serum is essentially similar in composition to plasma, but lacks fibrinogen and other factors that are consumed in the process of whole blood coagulation. Improper handling of whole blood results in hemolysis, giving a reddish tinge to serum.

Plasma is prepared by centrifugation (at 2000 x g for 10 min) of whole blood containing an anticoagulant. The sediment constitutes erythrocytes, while leucocytes separate as a buffy coat on the top. Plasma is a platelet rich fraction of whole blood.

Common anticoagulants used in the preparation of blood plasma are lithium salt of heparin (0.2 mg mL^{-1} or 12- 30 i.u mL^{-1}), ammonium / potassium oxalate (2 mg mL^{-1}), tri-sodium citrate (8 mg mL^{-1}) and ethylene diaminetetra acetic acid (EDTA) (1.2 - 2 mg mL^{-1}).

(ii) *Collection and preservation*: Blood is usually collected by venous puncture and kept in a glass tube containing sodium fluoride and potassium oxalate. For each mL of blood, 1 mg of sodium fluoride and 3 mg of potassium oxalate is added. Sodium fluoride inhibits glycolysis in RBC, while potassium oxalate acts as an anticoagulant.

The separated plasma or serum is best preserved at 4°C in the presence of added thimersol or sodium azide [0.05% (w/v)] as preservative.

Table 11.1 Body fluids and their methods of collection.

Type of body fluid	Method of collection
Whole blood	Vene puncture. Blood collected with anticoagulants (ex. heparin, EDTA). Arterial puncture (from radial, brachial or femoral artery).
Serum	Coagulated blood - centrifuged at 2500 x g. Supernatant - source of serum.
Plasma	Blood with anticoagulants, centrifuged at 2000 x g. Supernatant-source of plasma.
Cerebro-spinal fluid	By lumbar puncture (from sub-arachnoid region).
Serous fluid	By needle puncture to the serous space.
Mucosal secretions	Aspiration by Ryle's tube.
Sweat	Soaked onto filter paper.
Saliva	Chewing the adsorbent grade dental cotton for few minutes followed by centrifugal separation of saliva.
Urine	Directly passed into a clean glass or plastic containers (collected as - random, timed, 8-hour or 24-hour specimens). Refrigeration, freezing or addition of preservatives such as, thymol, acetic acid, toluene etc., to urine samples are needed to prevent undesirable changes during storage.

Quantitative Analysis of Blood Constituents

Experiment #1: Determination of glucose in serum/plasma by glucose oxidase-peroxidase method [§]

Principle: A convenient method of estimating glucose specifically in biological fluids is based on coupled enzyme assay. The first enzyme, glucose oxidase (a flavoprotein) catalyzes the oxidation of glucose to gluconic acid (δ -gluconolactone) with the liberation of hydrogen peroxide. The second enzyme, peroxidase (iron-porphyrin containing enzyme) catalyzes the oxidation of phenol, which combines with 4- aminoantipyrine to yield a pink chromogen (quinoneimine dye). The intensity of the colour developed is directly proportional to glucose concentration in the biological fluid. The pink coloured developed, is measured colorimetrically or spectrophotometrically at 505 nm. The sequence of enzymatic events is given below:

D - Glucose   GOD-FAD H_2O_2 POD-FeOH H_2O

GOD H_2O **POD**   Quinoneimine dye (Pink chromogen)

D - Gluconic acid   GOD-FADH$_2$ O_2 POD-FeOOH   4 - Aminoantipyrine + Phenol

1. D - Glucose $+ H_2O + O_2$ $\xrightarrow{\text{Glucose oxidase (GOD)}}$ D - Gluconic acid $+ H_2O_2$

2. H_2O_2 + 4 - Aminoantipyrine + Phenol $\xrightarrow{\text{Peroxidase (POD)}}$ Quinoneimine dye + H_2O (Pink chromogen)

Reagents: (i) Phosphate buffer (KH_2PO_4 & Na_2HPO_4), 100 mM, pH 7.0, containing 0.05% (w/v) thimersol as preservative (ii) Colour reagent (100 mL): Weigh and dissolve the following in the order, in 80 mL of phosphate buffer (100 mM, pH 7.0), 4-aminoantipyrine [(16 mg) , also known as 4-aminophenazone], glucose oxidase [1800 units (~1-1.2 mg solid); Sigma, G-7016, Type VII-S, source- fungal], peroxidase [100 units (~ 0.5 mg solid); Sigma, P-8250, Type II, source- horseradish], phenol (105 mg) and 0.01% (v/v) Tween-20. Mix and make up the volume with phosphate buffer to 100 mL in a volumetric flask. Store the reagent, refrigerated (4-8°C) in a brown bottle (reagent stable for two weeks).

Specimen: Serum or Plasma (EDTA or heparinzed). Prepare serum or plasma from blood specimen containing, potassium or sodium fluoride (1 mg mL^{-1} of blood).

Glucose standard (100 mg%): Weigh and dissolve 100 mg of dry D-glucose in 100 mL of distilled water saturated with benzoic acid (0.1% w/v), in a 100 mL volumetric flask.

Procedure:

(i)   Dispense the following into clean and dry test tubes labeled as Blank (**B**), Standard (**S**) and Test (**T**),

Protocol

	Reagents/Standard/Specimen	Blank	Standard	Test
1.	Colour reagent (mL)	3.0	3.0	3.0
2.	Phosphate buffer, 0.1 M, pH 7.0 (mL)	0.05	- -	- -
3.	Glucose standard (100 mg%) (mL)	- -	0.05	- -
4.	Specimen (serum/plasma) (mL)	- -	- -	0.05

Mix and incubate at 37°C for 15 min. in a constant temperature water bath.

Run the analysis in duplicates.

(ii) After 15 minutes, cool the tubes in ice cold water (2-4 °C) and immediately record the absorbance of standard and test, against blank, at 505 nm using a spectrophotometer.

(iii) Calculation: Calculate the concentration of glucose in the test sample using the following formula:

$$\text{Glucose (mg\%) or (mg/dL)} = \frac{\text{Absorbance of Test}}{\text{Absorbance of Standard}} \times 100$$

Note: (i) This method shows a good linearity in the range of 10 - 500 mg% of glucose. Thus, a calibration graph can also be generated by plotting the absorbance values of the standard against their respective concentrations in the range of 10-500 mg%. (ii) If the glucose concentration of the test sample exceeds 500 mg%, then dilute the sample (1:3 dilution) or appropriately in the range of detection. (iii) This method is also applicable for the analysis of glucose in CSF and urine specimens.

Workout

(i)   Repeat the experiment by using D - galactose as a test sample. Interpret the results.

(ii)   Analyse a urine specimen by this method.

§ (i) Tinder, P. Annals Clin. Biochem. 6: 24, (1969).
  (ii) Barham, D., & Tinder, P. Analyst, 97: 142, (1972).
  (iii) John, A. L., & Turner, K. Clin. Chem. 21: 1754, (1975).

Experiment #2: Determination of blood glucose by *Nelson-Somogyi* method §

Principle: The reducing sugars (glucose) in alkaline medium, upon heating form enediols which are good reducing agents. Enediols formed, reduces cupric (Cu^{2+}) copper (oxidising agent) to cuprous (Cu^+) state. The resulting cuprous oxide produced as red precipitate in the reaction reduces phosphomolybdic acid to yield a blue coloured complex (molybdenum blue), which is measured at 670 or 680 nm (λ_{max}).

Note: Substances like glucuronic acid, ascorbic acid, threonine, fructose and glutathione interfere with the assay (as all these compounds are reducing agents), resulting in erroneously high values.

Reagents: (i) Deproteinizing reagent: Zinc sulphate solution (5% w/v). Barium hydroxide solution 0.3 N. Weigh and transfer 15 g of barium hydroxide into a 1 L glass beaker containing 400 mL of distilled water and boil the contents for 5 - 7 minutes. Cool the contents and make up the volume to 500 mL and filter. For 5 mL of zinc sulphate solution, about 4.7 to 4.9 mL of barium hydroxide is required to neutralize. This can be ascertained by titrating zinc sulphate solution against barium hydroxide, in the presence of phenolphthalein indicator. The reaction of zinc sulphate with barium hydroxide results in the formation of zinc hydroxide (protein precipitating agent) and insoluble barium sulphate.

(ii) **Reagent - A**: Dissolve 25 g of anhydrous sodium carbonate, 25 g of potassium sodium tartrate, 20 g of sodium bicarbonate and 200 g of anhydrous sodium sulphate in 800 mL of distilled water. Warm and stir the contents in order to enhance the solubility. Cool the solution to room temperature and make up the volume to 1L with distilled water. Filter, if necessary. **Reagent - B**: Copper sulphate solution (15% w/v), containing two drops of conc. sulphuric acid. Alkaline copper-tartrate reagent: Mix 25 parts of reagent - A with 1 part of reagent - B. (this reagent should be prepared freshly). (iii) Arsenomolybdate reagent: It is prepared by dissolving 25 g of ammonium molybdate in 450 mL of distilled water. This is followed by addition of 21 mL of conc. sulphuric acid (added in small proportions), with constant stirring, using a glass rod. Add 3 g of disodium hydrogen arsenate, dissolved in 25 mL of water. Mix the reagent and transfer the contents into an amber coloured glass reagent bottle and incubate at 37°C, for 24 - 48 hours. Alternatively, the reagent can be prepared by heating the contents in a hot water bath at 55°C, for 30 - 40 minutes, instead of 48 hours incubation at 37°C.

Reference standard: Stock solution: Weigh 100 mg of analytical grade D-glucose and transfer into a 100 mL volumetric flask. Dissolve and make up the volume to 100 mL with 0.25 %(w/v) benzoic acid. Dilute 10 mL of the stock solution to 100 mL with distilled water to give a concentration of 100 µg/mL.

Procedure

(i)   To one millilitre of the standard solution (containing 10 -100 µg of glucose) or appropriately diluted unknown test sample or deproteinzed blood sample, add appropriate volume of distilled water to give a final volume of 2 mL. Add 2 mL of alkaline copper-tartrate reagent and vortex. Place the tubes in a boiling water bath, covered with glass marble, for 20 minutes. Cool the contents to room temperature and add 1 mL of arsenomolybdate reagent and mix. Make up the volume to 10 mL by adding distilled water. Record the absorbance of the solutions at 670 or 680 nm against a reagent blank.

(ii)   Construct a calibration curve on a graph paper, by plotting the concentration of glucose (10 -100 µg) on *x*- axis and absorbance at 670 or 680 nm, on the *y*- axis. Compute the concentration of glucose in the test sample from the calibration plot.

Protocol

S.No	Std. Glucose (mL)	Distilled water (mL)	Conc. of Glucose (μg)	Alkaline Copper reagent (mL)	Arseno-molybdate reagent (mL)	Distilled water (mL)	Absorbance at 670/680 nm
1.	0.1	1.9	10	2.0	1.0	5.0	
2.	0.2	1.8	20	"	"	"	
3.	0.3	1.7	30	"	"	"	
4.	0.4	1.6	40	"	"	"	
5.	0.5	1.5	50	"	"	"	
6.	0.6	1.4	60	"	"	"	
7.	0.7	1.3	70	"	"	"	
8.	0.8	1.2	80	"	"	"	
9.	0.9	1.1	90	"	"	"	
10.	1.0	1.0	100	"	"	"	
11.	Blank	2.0	–	"	"	"	
12.	Sample		To be determined	"	"	"	

" = same volume

Clinical Significance

Hyperglycaemia: Hyperglycaemia is a clinical condition wherein the blood glucose level exceeds 120 mg/dL during fasting conditions, the normal range being 90-120 mg/dL. Increased blood glucose concentration is observed in disorders such as *Diabetes mellitus*, hyperactivity of the thyroid, the pituitary, the adrenal cortex and the adrenal medulla. A value greater than 200 mg/dL of plasma glucose by glucose tolerance test (GTT), at 2 hours is a diagnostic indicator of diabetes. A plasma glucose value ranging from 140 -200 mg/dL is an indication of impaired glucose tolerance (IGT). It should be noted that, subjects with IGT are not considered as normal and are to be followed up for diabetes at a later period of time. Increase in blood glucose levels are also observed in sepsis and in infectious diseases such as *meningitis, encephalitis*.

Hypoglycaemia: When blood glucose level is less than 50 mg/dL, such clinical condition is referred to as hypoglycaemia. The causes of this clinical status may be resultant of more than one factor, such as (i) hypofunctioning of the thyroid, the pituitary or the adrenal cortex (ii) starvation (iii) diseases of glycogen storage (type I, III & VI) (iv) overdose of insulin during treatment of diabetes and (v) drug induced hypoglycaemia. Hypoglycaemic condition is also observed in *Addison's* disease.

Blood and plasma sugar levels in normal and abnormal conditions

Condition	Glucose concentration (mg/dL)*	
	Blood	Plasma
Normal - Fasting	60 - 90	75 - 105
Normal -Post-prandial	90 - 120	100 - 140
Diabetic - Fasting	**> 110**	**> 140**
Impaired glucose tolerance (2h)		140 - 199
Glucose tolerance test (2h)		> 200

Note: * The concentration of glucose in plasma is 10 -15% higher than whole blood

Workout: Determine the glucose content in (i) human blood sample (a) *fasting blood sugar* - blood sample collected after 12 hours of fasting; (b) *post- prandial blood sugar* - blood sample collected 2 hours after the consumption of carbohydrate rich meal (c) *random blood sugar* - blood collected at any time (ii) human urine sample (24 hour urine).

Note: *Blood sample preparation*: To 0.1 mL of fresh blood sample, add 3.5 mL of distilled water, followed by addition of 0.2 mL of 5% zinc sulphate solution and 0.2 mL of barium hydroxide (0.3 N) and vortex. Centrifuge the contents at 2500 rpm for 10 minutes. Collect the supernatant into a clean test tube. Estimate the sugar content in the supernatant by the above method, using aliquots of 1 and 2 mL. Calculate the blood glucose concentration either by using the calibration plot or by the following equation,

Calculation of blood glucose (mg/dL):

$$\frac{\text{Absorbance of test sample}}{\text{Absorbance of standard}} \times \frac{\text{Volume of Blood (100 mL)}}{\text{Volume of blood sample taken (0.025 or 0.05 mL)}}$$

$$\times \text{ Conc. of glucose std.}$$

Urinary glucose estimation: Take an aliquot of 10 mL from the 24 h urine sample and centrifuge at 2500 rpm for 10 minutes. Appropriately, dilute 1 mL of the clarified urine sample (1:5, 1:10 , 1: 20 etc.,) for sugar estimation. Compute the concentration of urinary glucose form the standard calibration plot.

Urinary glucose can also be determined by titrimetric method as described in **7. *Quantitative analysis***.

^{s}(i)   Nelson, N., J. Biol. Chem. 153: 375, (1944).
(ii)   Somogyi, M. J. Biol. Chem. 195:19, (1952).

Experiment #3: Determination of haemoglobin by cyanmethaemoglobin method [s]

Principle: Haemoglobin in the presence of potassium ferricyanide at alkaline pH is oxidised to methaemoglobin (*iron in ferric state*). The methaemoglobin formed, reacts with potassium cyanide to yield a red coloured complex, namely cyanmethaemoglobin, which is photometrically measured at 540 nm. This procedure is an internationally approved and widely used reference method for the determination of blood haemoglobin.

Reagents: *Drabkin's* solution: Dissolve 50 mg of potassium cyanide, 20 mg of potassium ferricyanide and 1g of sodium bicarbonate in fresh glass distilled water and make up the volume to 1L in a volumetric flask. Store the reagent in an amber coloured reagent bottle.

> **Caution**: Potassium cyanide is a highly toxic chemical. It is preferable to prepare this reagent under the supervision of an instructor. Avoid contact with skin and inhalation while handling this chemical. This chemical should be handled in a chemical safety hood, wearing disposable gloves. (This reagent can be procured commercially also).

Standard haemoglobin solution: Prepare 61 mg% haemoglobin (Hb) solution as source of reference standard.

Procedure: Dispense 5 mL each of *Drabkin's* solution into three test tubes labelled as reagent blank, standard and test solution, by the aid of pro-pipette/auto-dispenser (do not mouth pipette this reagent). Add 0.02mL of distilled water into blank and 0.02 mL of standard and whole blood (unclotted blood or directly from the finger puncture) into the respective test tubes. Mix and leave the contents at room temperature for 5 min. Read the absorbance of the standard and the test sample immediately, at 540 nm against the reagent blank in a colorimeter.

Calculation: The content of haemoglobin in the test sample is calculated by the following equation:

$$\text{Blood haemoglobin g \%} = \frac{\text{Absorbance of test sample}}{\text{Absorbance of standard}} \times \frac{61 \text{ (Conc. of Hb)} \times 251 \text{ (dilution factor)}}{1000}$$

Clinical Significance

Blood haemoglobin level is a good index in the assessment of anaemia. Clinically, anaemia is defined as a condition wherein there is a reduction in the circulating levels of haemoglobin. The contributory causes for the development of anaemia include, failure of erythrocyte production or destruction of red blood cells or excessive loss of blood.

Blood haemoglobin levels in normal and anaemic conditions.

Group	Normal range (Hb g %)	Anaemia (Hb g %)*
Adult men	13.5 - 18.0	< 12.0
Adult women	11.5 - 16.4	< 12.0
Children (1-10 years)	11.2 - 12.0	< 11.0

* World Health Organization Technical Report Series # 503. WHO, Geneva, 1972.

Workout: Analyse haemoglobin content in a clinical sample.

§ Van Kampen E. J., & Zijlstra, W. G., Clin. Chim. Acta. 6: 538, (1961).

Experiment #4: Estimation of total serum cholesterol by *Zak & Henly's* method §

Principle: The chemical property of cholesterol to undergo dehydration, oxidation and sulphonation with chemical reagents such as sulphuric acid, acetic acid and acetic anhydride, to a red coloured complex in the presence of ferric ion (ferric chloride), forms the basis for the quantitation of total serum cholesterol. The red colour has a λ_{max} at 560 nm.

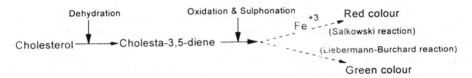

Reagents: (i) Glacial acetic acid (ii) Concentrated sulphuric acid (iii) Ferric chloride reagent (*Kiliani's* reagent): Prepare 8 % (w/v) ferric chloride monohydrate in glacial acetic acid. Dilute 1mL of this reagent to 100 mL with glacial acetic before use.

Cholesterol standard: Weigh 100 mg of pure cholesterol and transfer into a clean 100 mL volumetric flask. Dissolve and make up the volume to 100 mL in glacial acetic acid.

Procedure

(i)   Dispense 9.9 mL of ferric chloride reagent into a stoppered and graduated centrifuge tube, using a pro-pipette. Add 0.1 mL of the test serum sample, (perform duplicate analysis) and mix thoroughly. Allow it to stand for 15 - 20 min with intermittent mixing. Centrifuge the contents at room temperature for 15 min at 2500 rpm. Transfer 5 mL of the clear supernatant into a clean test tube. Add 3 mL of conc. sulphuric acid using a burette and mix the contents thoroughly. After 30 min record the absorbance at 560 nm against reagent blank. Simultaneously, run a set of standards as given below:

Protocol

S.No	Standard Cholesterol (mL)	Conc. of Cholesterol (mg)	Ferric chloride (mL)	Conc. H_2SO_4 (mL)	Absorbance at 560 nm
1.	Blank	- -	5.0	3.0	
2.	0.05	0.05	4.95	"	
3.	0.1	0.1	4.90	"	
4.	0.2	0.2	4.80	"	
5.	0.3	0.3	4.70	"	
6.	0.4	0.4	4.60	"	
7.	Test (5.0 mL)*	To be determined	- -	"	

* 5.0mL of the supernatant obtained from serum treated with ferric chloride reagent.

(ii) Construct a calibration curve on a graph paper, by plotting the concentration of cholesterol (0.05 - 0.4 mg) on *x* axis and absorbance at 560 nm, on the *y* axis. Compute the concentration of cholesterol in the test sample from the calibration plot. Use dilution factor of 2, while computing the concentration of cholesterol in the test sample. Report the cholesterol concentration of the test sample as mg/dL.

Clinical Significance

Increased serum levels of cholesterol have been positively correlated to the development of *atherosclerosis* leading to coronary heart disease (CHD). Hypercholesterolaemia is also associated in clinical conditions such as, nephrosis, obstructive jaundice, *Diabetes mellitus* and *Myxoedema*. On the contrary, decreased serum cholesterol levels are observed in hyperthyroidism, hepatocellular injury and intestinal obstruction. Total blood cholesterol levels in adults range from 180 - 240 mg/dL

Risk of CHD associated with increased serum cholesterol levels

Group	Total serum cholesterol [mg(w/v)% or mg/dL]		
	Desirable	**Marginal risk**	**Potential risk**
Adults	< 200	201 - 239	> 240
Children & Adolescents	< 170	171 - 199	> 200

Workout: Compare the normal serum cholesterol content of (i) Adult and (ii) Adolescent age groups.

§ Zak, B., Dickenbaum, R. L., White, E. G., Burnett, H., & Cherney, P. J. Amer. J. Clin. Pathol. 24: 1307, (1954).

Experiment #5: Determination of serum bilirubin by *van den Bergh* reaction §

Principle: Bilirubin is an orange-yellow pigment derived from heme catabolism. It exists in conjugated and unconjugated forms in blood. The quantitation of bilirubin in serum is based on the formation of a reddish-purple coloured dipyrrole azo-derivative at neutral pH, a product formed by the reaction of diazotized sulphanilic acid with bilirubin. This gives an estimate of conjugated bilirubin (direct reaction). If diazo-reaction is carried out in the presence of alcohol, as an accelerator, it gives a measure of unconjugated (indirect reaction) bilirubin. The colour is measured photometrically at 540 nm.

Dipyrrole azo-derivative (reddish-purple)

Reagents: (i) Diazo-reagent: *Reagent A*: Dissolve 1 g of sulphanilic acid in 15 mL of conc. HCl and make up the volume to 1L with distilled water. *Reagent B*: Prepare 0.5% (w/v) sodium nitrite ($NaNO_2$) solution. Diazo-reagent should be prepared freshly before use, by adding 0.8 mL of reagent B with 10 mL of reagent A. (ii) prepare 0.18 N HCl (iii) methanol (iv) chloroform.

Standard bilirubin: Weigh 10 mg of pure bilirubin and transfer it into a 100 mL volumetric flask. Dissolve and make up the volume with chloroform.

Dilute 1 mL of the stock to 50 mL in a volumetric flask with methanol to give a working standard of 2 µg/mL.

Procedure

(i) Dispense 0.2 mL of the test serum (in duplicates) into labelled test tubes. Add, 1.8 mL of distille.! water and 0.5 mL of diazo-reagent using a pro-pipette, followed by addition of 2.5 mL of methanol. Mix the contents and allow the colour to develop for 30 min, at room temperature. Record the absorbance against blank at 540 nm. Simultaneously, run a set of standards as given in the following protocol:

Protocol

S.No	Standard Bilirubin (mL)	Conc. of Bilirubin (µg)	Diazo-reagent (mL)	Methanol (mL)	Distilled water (mL)	HCl (0.18 N) (mL)	Absorbance at 540 nm
1.	Blank	- -	- -	2.5	2.0	0.5	
2.	0.5	1.0	0.5	4.0	- -	- -	
3.	1.0	2.0	"	3.5	- -	- -	
4.	2.0	4.0	"	2.5	- -	- -	
5.	3.0	6.0	"	1.5	- -	- -	
6.	4.0	8.0	"	0.5	- -	- -	

" = same volume; - - = No addition

Specimen analysis:

S.No.	Sample	Conc. of Bilirubin	Distilled water (mL)	Diazo-reagent (mL)	Methanol (mL)	Absorbance at 540 nm
7.	Test serum 0.2mL	To be determined	1.8	0.5	2.5	

(ii) Construct a calibration curve on a graph paper, by plotting the concentration of bilirubin (1.0 - 8.0 µg) on *x*- axis and absorbance at

540 nm, on the *y*- axis. Compute the concentration of bilirubin in the test specimen from the calibration plot. Express the bilirubin concentration in the test sample as mg/dL.

Note: Following protocol is used to determine the concentration of serum conjugated bilirubin (bilirubin biglucuronide). The reaction is performed in the absence of methanol for 1 min and immediately the absorbance is measured against the blank. Calculate the concentration of bilirubin from the calibration plot.

S.No	Sample	Conc. of conjugated Bilirubin	Distilled water (mL)	Diazo-reagent (mL)	Absorbance at 540 nm
8.	Test serum 0.2 mL	To be determined	4.3	0.5	

Calculation: Total serum bilirubin (mg/dL) = (Conc. of unconjugated + conjugated bilirubin).

Clinical Significance

Elevated serum bilirubin levels are found in hepatocellular and obstructive jaundice, with failure of excretion into the bile. Jaundice is classified into two major types namely, *retention* and *regurgitation*. The characteristic feature of retention jaundice is elevated levels of unconjugated bilirubin in circulation, which is caused by (i) excessive haemolysis (ii) defective hepatic uptake and (iii) decreased hepatic conjugation activity.

Regurgitation jaundice is characterized by an increased plasma level of conjugated bilirubin due to impaired hepatic excretion or cholestasis. Obstruction of biliary channel due to *atresia*, gallstones, bacterial infections, tumours etc., lead to elevation of serum conjugated bilirubin by its regurgitation into circulation. Acute hepatitis due to ingestion of toxins, certain drugs and viral infection results in excretion of bilirubin in urine and decrease in the excretion of faecal urobilinogen. Familial jaundice, such as *Crigler-Najjar* syndrome, *Gilbert's* syndrome, *Lucey-Driscoll* syndrome, *Rotor* syndrome is characterized by abnormal serum bilirubin levels.

Normal and abnormal serum bilirubin levels

Normal	(Range - mg/dL)
Total	0.3 - 1.0
Conjugated	0.1 - 0.3
Unconjugated	0.2 - 0.7
Abnormal	
Haemolytic anaemia's (unconjugated)	<4.0
Pernicious anaemia (unconjugated)	< 3.0
Retention jaundice	1.0 - 20
Regurgitation jaundice	20 - 50

Workout: Analyse the bilirubin content in a specimen of clinical origin.

§ Malloy, H. T., & Evelyn, K. A. J. Biol. Chem. 119: 481, (1937).

Experiment #6: Determination of serum uric acid *§*

Principle: Serum uric acid (2, 6, 8 -trioxypurine) is an end product of purine catabolism in humans. Uric acid reduces colourless phosphotungstic acid to a blue coloured chromogen - tungsten blue, at alkaline pH. The colour formed is colorimetrically determined at 700 nm.

Note: Endogenous serum constituents such as Vit. C, glutathione and cysteine interfere in this reaction.

Reagents: (i) Sodium tungstate (10% w/v) (ii) Sulphuric acid (0.67 N) (iii) Sodium carbonate (10%w/v) (iv) Phosphotungstic acid reagent (stock solution): Weigh 50 g of sodium tungstate and transfer into a 1 L round bottom flask. Dissolve sodium tungstate by adding 400 mL of distilled water. Later, add 40 mL of ortho-phosphoric acid (85-88%) and reflux the contents for 2 hours, using an electric heating mantle. To avoid bumping, add few porcelain chips. Cool to room temperature, transfer the contents into a 500 mL volumetric flask and make up the volume with distilled water. Store the reagent in an amber coloured reagent bottle. This reagent is stable for several months. Before use, dilute 1 mL of the stock reagent to 10 mL with distilled water (v) Lithium carbonate solution (0.3% w/v).

Standard uric acid: Weigh 100 mg of pure uric acid and transfer it into a 100 mL beaker. Add 20 mL of lithium carbonate solution (0.3 % w/v) and warm the contents to 60°C in a hot water bath, until uric acid dissolves. Stir the contents intermittently with the aid of a glass rod. Cool to room temperature and transfer the contents quantitatively into a 100 mL volumetric flask. Add 2 mL of formalin solution, followed by 1 mL of 1:1 diluted glacial acetic acid solution. Mix and make up the volume to 100 mL with distilled water. Working standard: Dilute 0.5 mL of the stock to 100 mL with distilled water to give a concentration of 5 μg/mL.

Procedure

(i)   Transfer 1 mL of test serum into a clean graduated centrifuge tube (15 mL capacity). Add 7 mL of distilled water, followed by addition of 1 mL each of sodium tungstate (10 %) and sulphuric acid (0.67 N). Mix the contents gently and allow it to stand for 10 -15 min. Centrifuge the contents for 15 min at 2500 rpm. Use 5 mL of the clear supernatant for uric acid analysis.

(ii)   Add 1mL each of sodium carbonate (10%) and phosphotungstic acid to (a) Blank (b) Standard and (iii) Test, as given below. Mix the contents and measure the colour photometrically at 700 nm.

Protocol:

S.No	Experiment	Sodium carbonate (10%) (mL)	Phosphotungstic acid reagent (mL)	Absorbance at 700 nm
1.	Blank (5 mL distilled water)	1.0	1.0	
2.	Standard uric acid (5 mL , Conc. 25μg)	1.0	1.0	
3.	Test (5 mL of deproteinated serum supernatant)	1.0	1.0	

(iii) **Calculation**

$$\text{Serum uric acid (mg/dL)} = \frac{x}{y} \times \frac{100 \times 25}{0.5 \times 1000} = \frac{x}{y} \times 5$$

Where, x = Absorbance value of test sample; y = Absorbance value of the uric acid standard.

Clinical Significance

In clinical conditions wherein, the uric acid concentration exceeds 8 mg/dL is normally referred to as hyperuriciemia. Various diseased states, such as renal failure, *nephrolithiasis*, *polycythemia*, chronic nephritic leukemia and gout, influence the clinical condition of *hyperuriciemia*. Gout is a multi-factorial syndrome, wherein (i) there is an increase in serum urate concentration (ii) accumulation of monosodium urate monohydrate crystals in leukocytes of synovial fluid (iii) aggregation and deposition of monosodium urate monohydrate crystals in and around joints (referred as *tophi*) resulting in arthritic condition. In acute case of gout, plasma urate level increase to a maximum of 15 mg/dL.

Liver disorders such as cirrhosis lead to hypouriciemia, wherein the blood uric acid level of < 2 mg/dL has been recorded, as liver is the major site for uric acid synthesis. Normal plasma uric acid level in men (2 - 7 mg/dL) is higher than women (2- 6 mg/dL).

Workout: Analyse uric acid in a specimen of clinical origin.

§ Carway, W. T. Amer. J. Clin. Pathol. 25: 840, (1955).

Experiment #7: Determination of blood urea (diacetylmonoxime method) §

Principle: Urea, a major nitrogenous compound derived from protein metabolism, is synthesized in liver and excreted through kidneys. It constitutes 80-90% of the body nitrogen excreted. Blood urea nitrogen levels alter with dietary protein intake. Urea when heated with compounds containing adjacent ketonic groups- like diacetyl monoxime, undergo condensation to a purple coloured diazine derivative, at acidic pH, which is photometrically measured at 520 nm. The presence of thiosemicarbazide and ferric ions catalyze the enhancement of the colour intensity in the reaction. This method is free from interference with compounds such as ammonia or acetone.

Diacetylmonoxime → Diacetyl (2,3- Butadione) → Purple coloured Diazine derivative

Reagents: (i) Trichloroacetic acid (10% w/v) (ii) Diacetylmonoxime solution (2.5% w/v) (iii) Thiosemicarbazide solution (2.5% w/v) (iv) Ferric chloride solution (5% w/v) (v) Acidified ferric chloride solution: Add and mix 1 mL of conc. H_2SO_4 to 99 mL of 5% ferric chloride solution (vi) Acid reagent: It is prepared by mixing 10 mL of ortho-phosphoric acid and 80 mL of conc. H_2SO_4 and 10 mL of acidified ferric chloride reagent. The volume is made up to 1 L with distilled water in a volumetric flask (vii) Colour reagent: This is prepared by mixing 30 mL of acid reagent, 200 mL of distilled water, 10 mL of 2.5 % diacetylmonoxime solution and 2.5 mL of thiosemicarbazide solution. This reagent is stored in an amber coloured reagent bottle (stable for 6 months, if refrigerated).

Standard urea: Weigh and dissolve 30 mg urea in distilled water. Make up the volume to 100 mL in a volumetric flask.

Procedure:

(i)   Take 0.2 mL of whole blood or serum in a centrifuge tube and add 0.8 mL of normal saline, followed by addition of 1 mL of 10 % TCA. Mix and centrifuge at 2500 rpm for 15 min. Transfer 0.5 mL of the supernatant into a clean test tube labelled as test. Run a blank and a set of urea standards as given below. Add 3 mL of the colour reagent to blank, urea standards, test and heat the test tubes in a boiling water bath for 20 min. Cool the tubes to room temperature and measure the colour developed within 15 minutes against blank at 520 nm using a photoelectric colorimeter.

Protocol

S.No	Standard Urea (mL)	Conc. of Urea (µg)	Distilled water (mL)	Colour reagent (mL)	Absorbance at 520 nm
1.	Blank	- -	1.0	3.0	
2.	0.1	30	0.9	"	
3.	0.2	60	0.8	"	
4.	0.3	90	0.7	"	
5.	0.5	150	0.5	"	
6.	0.8	240	0.2	"	
7.	1.0	300	- -	"	
8.	Test (0.5 mL* of deproteinated blood or serum supernatant)	To be determined	0.5	"	

" = same volume; - - = No addition; * or appropriate volume; Run test in duplicates

(ii)   Construct a calibration curve on a graph paper, by plotting the concentration of urea on x- axis and absorbance at 520 nm, on the y-axis. Compute the concentration of urea in the test specimen from the calibration plot. Express the concentration of blood urea in the test sample as mg/dL. Use dilution factor, while calculating the urea concentration.

Clinical significance

Any clinical condition which is associated with imbalance in the physiological function of liver and kidneys lead to abnormal blood urea levels. Normal blood urea level depends upon the balance between its hepatic synthesis and subsequent renal excretion. Major clinical conditions that result in altered urea levels include, renal dysfunction and renal diseases which primarily affect the functioning of the kidneys.

Renal dysfunction: Pre-renal causes that result in blood urea retention are attributable to reduction in renal blood flow and glomerular filtration rate. Under this category, clinical conditions such as, acute gastroenteritis, pyloric-stenosis, upper gastrointestinal haemorrhage, burns, toxaemia, diabetic ketoacidosis, haemolytic anaemia and ulcerative colitis are included.

Renal diseases: Renal diseases that cause retention of blood urea include acute & chronic *glomerulonephritis, polyarteritis nodosa*, and bilateral renal tuberculosis. Also, chronic vascular diseases such as benign and malignant hypertension and renal vein thrombosis result in high blood urea levels. Further, obstruction to flow of urine due to urethral blockade (by stones or prostatic enlargement) leads to bilateral hydronephrosis causing blood urea retention.

Hepatic diseases: Severe hepatic diseases such as liver cirrhosis and acute hepatic necrosis cause low blood urea levels due to depressed urea synthesis. Significant reduction of blood urea nitrogen is indicative of liver dysfunction. Other factors that influence blood urea levels include, high protein diet, increased protein catabolism, starvation and muscle wasting.

Blood urea levels (mg/dL) in normal & abnormal conditions

Normal*	14 - 50
Abnormal:	
(i) Diabetic coma	50 - 150
(ii) Pyloric-stenosis	> 200
(iii) Acute glomerulonephritis & ulcerative colitis	> 300
(iv) Malignant hypertension & chronic pyelonephritis	> 420

* Blood urea levels are higher in men than in women

Workout: Determine the urea concentration in a urine specimen and compare it with blood.

§ Marsh, W. H., Fingerhut, B., & Miller, H. Clin. Chem. 11: 624, (1965).

Experiment 8: Determination of albumin : globulin ratio (A/G ratio) [§]

(i) Determination of Albumin by dye binding method:

Principle: Albumin, the principle plasma protein exists as a positively charged species below its isoelectric point (pH_I 4.8). The positively charged albumin exhibits a strong affinity for anionic dyes such as methyl orange, bromocresol green (BCG) and bromophenol blue. This protein-dye interaction forms the basis for quantitation of plasma/serum albumin. Amongst these dyes, BCG is found to be more suitable and sensitive for albumin determination. Bromocresol green exists as a blue divalent anion at neutral or alkaline pH, while in acidic pH, below its pk_a value (4.7), the dye is present in an undissociated form (yellow colour). The binding of the dye to albumin at acidic pH upsets this equilibrium of undissociated and dissociated forms, resulting in increased absorbance at 630 nm.

Unionized form of BCG (yellow)   Ionized form of BCG (blue)

Ionization of dye

Reagents: (i) Dye solution: Weigh and suspend 0.7 g of BCG dye in 30 - 40 mL of distilled water. Add 1 mL of 1N NaOH and mix to dissolve the dye. Make up the volume to 100 mL with distilled water in a volumetric flask. (ii) Brij-35 (23 - lauryl ether) solution (25% w/v): Warm 12.5 g of Brij-35 in 20 mL of distilled water and make up the volume to 50 mL with distilled water. Buffered dye reagent: This is prepared by dissolving 5.9 g of succinic acid, 1 g of sodium hydroxide in 500 mL of distilled water. Adjust the pH to 4.1 - 4.2. To this add 2.5 mL of Brij-35 solution, followed by addition of 100 mg of sodium azide. Add 8 mL of the dye solution, mix and check the pH (adjust to 4.1 - 4.2 if needed) of the reagent. Make up the volume to 1 L with distilled water. The reagent is stable for 6 months, if refrigerated.

Albumin standard: Prepare 2, 3, 4 and 6 g% of human serum albumin (HSA) in 0.1 N NaOH.

Note: Do not use bovine or equine serum albumin as reference standard, as their interaction with the dye is different form that of HSA.

Procedure

(i)   To 20 µL of the test serum, add 4 mL of the buffered dye reagent, mix and allow it to stand at 25⁰C for 10 min. Measure the absorbance immediately, against reagent blank at 630 nm. Run a set of human serum albumin standards (2 - 6 g%) along with the test sample. If the concentration of standard HSA and absorbance show a linear relationship, then use 4 g% absorbance value, for computing the concentration of albumin in test serum sample.

Protocol

S.No.	Standard HSA (mL)	Concentration of HSA (µg)	Buffered dye reagent (mL)	Absorbance at 630 nm
1.	Blank#	- -	4 .0	
2.	0.02	400	"	
3.	0.02	600	"	
4.	0.02	800	"	
5.	0.02	1200	"	
6.	Test sample (0.02)*	To be determined	"	

* Run in duplicates; " = same volume; # - add 0.02 mL of distilled water

Calculation: Serum albumin (g/dL) = $\dfrac{\text{Absorbance of test sample}}{\text{Absorbance of standard HSA}} \times 4$

(ii) Determination of total serum proteins. Determine the concentration of total serum proteins in the test sample by Biuret method, as detailed in *7. Quantitative analysis*.

(iii) Determination of serum globulin content: This is calculated by subtracting the albumin content from total serum proteins,

Serum globulin content (g%)] =

[Total serum protein content (g%) - Serum albumin content (g%)]
(As determined by Biuret method)   (As determined by dye binding method)

(iv) Calculate A/G ratio (Albumin:Globulin ratio) from the experimental data.

Clinical Significance

A/G ratio falls in clinical conditions such as liver cirrhosis and nephrotic syndrome resulting in hypoproteinemia. Lowered serum albumin levels leads to edema and alterations in blood calcium levels. Decreased synthesis of albumin is observed in malnutrition, malabsorption and chronic hepatic diseases. Fever and sepsis cause increased catabolism of serum albumin. Burns and renal disorders lead to increased loss of serum albumin. Conversely, hyperproteinemia, a condition wherein there is an increase in total protein concentration in serum, is observed in diseases such as *Kalazar,* multiple myeloma. In these subjects, there is an absolute increase in the globulin concentration, thus resulting in the fall of A/G ratio.

Normal values for serum proteins, albumin, globulin and A/G ratio

Total Serum Proteins	6 - 8 g/dL
Albumin	3.5 - 5.5 g/dL
Globulin	2 - 4 g/dL
A/G (ratio)	Range 1.5 - 2.5

Workout: Determine the A:G ratio in a clinical blood sample.

$ Spencer, K., & Price, C. P. Ann. Clin. Biochem. 14: 105, (1977).

B. CLINICAL ENZYMOLOGY

Measurement of enzymes that are indicative of cellular dysfunction or damage is of diagnostic value in the clinical assessment of the diseased state. Both, physiological and pathological states influence the blood enzyme level(s) or altered activity(ies). Elevated levels of serum/plasma enzyme(s) may be resultant of leakage from the necrotic or damaged cells/tissues/ organs due to (i) ischemia (ii) neoplasia (iii) carcinoma (iv) decrease in their normal clearance and (v) impairment in the normal process of excretion, as seen in biliary obstruction. Application of optimized and standardized enzyme assay conditions is a perquisite for the assessment of clinically important diagnostic enzymes. This aspect of analysis is a major component of clinical enzymology. Table 11.2 shows some clinically important diagnostic enzymes.

Table 11.2 Some clinically important enzymes and their diagnostic value.

Enzyme	Source	Diagnostic application
Acid phosphatase (ACP)	Prostate	Prostate carcinoma, myelocytic leukemia, metastatic bone disease
α-Amylase	Saliva & pancreatic juice	Acute pancreatitis, perforated duodenal ulcer, peritonitis, cholecystitis.
Alkaline phosphatase (ALP)	Bone, liver, intestinal mucosa, kidneys and placenta.	Hepatobiliary disease, bone diseases, liver & bone malignancies.
Alanine amino-transferase (ALT)	Liver, heart, skeletal muscles	Hepatocellular injury or damage.
Aspartate amino-transferase (AST)	Liver, heart, skeletal muscles, kidneys, erythrocytes	Myocardial infarction, muscle and hepato-cellular disorders, severe hemolysis of erythrocytes.
Creatine kinase (CK)	Heart, skeletal and smooth muscles, brain	Myocardial infarction and muscle diseases, disorders of central nervous system, malignancies.
γ- Glutamyl transferase (GGT)	Liver & kidneys	Hepato-biliary disease, alcoholism, malignancies.
Lactate dehydro-genase (LDH)	Skeletal muscles, heart, liver, lymph nodes.	Myocardial infarction, hepato-cellular disorders, hematological disorders, skeletal muscle diseases, malignancies.

Experiment #9: **Assay of serum alkaline phosphatase** (EC 3.1.3.1, *o*-phosphoric monoester phosphohydrolase, *p*H optima alkaline)

Serum alkaline phosphatase can be assayed by using a variety of substrates. These include, β-glycerophosphate, phenyl phosphate, phenolphthalein phosphate and *p*-nitrophenyl phosphate (*p*-NPP). In each case the rate of phosphorolysis is measured photometrically.

Principle: Serum alkaline phosphatase catalyses the conversion of the substrate, *p*-nitrophenyl phosphate to *p*-nitrophenol (a yellow coloured product) and inorganic phosphate, at alkaline *p*H. The enzyme activity is measured by recording the absorbance due to the formation of *p*-nitrophenol, at 410 nm.

p-nitrophenyl phosphate   *p*-nitrophenol   Inorganic phosphate

Reagents: (i) Buffered substrate reagent: Prepare 1.2%(v/v) diethanolamine-HCl buffer containing 200 mg of *p*-nitrophenyl phosphate and 0.05 mM MgCl$_2$, *p*H 9.6. (ii) NaOH (0.05 N) (iii) Conc. HCl.

Stock solution: Weigh 139 mg of *p*-nitrophenol (*p*-NP) and transfer into a 100 mL volumetric flask. Dissolve and make up the volume to 100 mL with 0.05 N NaOH. This gives a concentration of 10 μmoles of *p*-NP mL^{-1}.

Working range: Dilute 5 mL of the stock to 100 mL with 0.05 N NaOH, to give a concentration of 0.5 μmole mL^{-1}. Working range of *p*-NP - 0.05 to 0.5 μmoles.

Procedure

(i) Pipette out 1 ml of buffered substrate in three test tubes labelled as reagent blank (RB), serum blank (SB) and test serum (T) and warm the tubes at 37°C in a constant temperature water bath for few minutes. Add 0.1 mL conc. HCl to SB, followed by addition of 0.05 mL of serum to SB and T. Note the time. Gently shake and incubate the test tubes exactly for 30 min at 37°C in a constant temperature water bath, along with reagent blank.

Protocol*

Volume in mL	Reagent Blank (RB)	Serum Blank (SB)	Test sample (T)
1. Buffered substrate	1.0	1.0	1.0
2. Serum	- -	0.05	0.05
3. Alkali (0.05 N)	1.95	1.85	1.95
4. HCl	- -	0.1**	- -
5. Distilled water	0.05	- -	- -

* Run the assay in duplicate. ** Add conc. HCl, before the addition of serum

(ii) Add alkali (0.05 N) to stop the enzymatic reaction to all the test tubes as given in the above table. Make the final volume to 3 mL by addition of distilled water. Mix and record the absorbance at 410 nm against the reagent blank.

(iii) Record the absorbance of serum blank against the reagent blank. Deduct this value from test sample. This takes into account the colour due to serum only.

Calibration plot for standard *p*-nitrophenol: Pipette out 0.1 to 1 mL (0.1, 0.2, 0.4, 0.8 and 1.0 mL) of the working standard into separate labelled test tubes. Make up the volume to 3 mL with 0.05 N NaOH and mix. Use 0.05 N NaOH as reagent blank. Measure the absorbance of the standards against blank at 410 nm. Construct a calibration curve by plotting conc. of *p*-nitrophenol in micromoles on the *x* -axis and the absorbance on *y*- axis.

(iv) Calculation of enzyme activity: Determine the μmoles of *p*-nitrophenol formed due to enzyme activity of serum (0.05 mL), using the calibration plot. Calculate the units of enzyme activity in terms of μmoles of *p*-NP formed per minute per liter of serum.

$$\text{Enzyme activity (UL}^{-1}) = \text{μmoles of } p\text{-NP formed} \times \frac{1000 \text{ mL}}{0.05 \text{ mL}} \times \frac{1}{30}$$

Clinical Significance

Alkaline phosphatase (ALP) is one of the widely distributed enzymes in the body with a preponderant occurrence in bone, liver, kidneys, gut mucosa and placenta. This enzyme exists in multiple forms. In liver this enzyme is membrane bound and is present in bile canaliculi and sinusoidal surfaces of the hepatocytes. Increased activity is observed in hepatocellular disease and cholestasis. In hepatitis, the raise in the activity is 2 to 3 folds, than the normal. In biliary obstruction, there is a substantial increase by 20 folds, as compared to normal value. Thus, ALP is a diagnostic marker for biliary obstruction. In addition, alkaline phosphatase is also a useful diagnostic enzyme in the assessment of bone diseases such as *Paget's* disease, osteomalacia and bone tumor. Normal serum levels of this enzyme in adult male and female are 38-94 and 28-111 UL^{-1}, respectively.

Experiment #10: Assay of serum alanine aminotransferase (ALT), earlier known as serum glutamate-pyruvate transminase (SGPT) (EC 2.6.1.2)§

Principle: The determination of serum alanine aminotransferase activity is based on coupled enzyme assay. This enzyme catalyses the transfer of amino group from alanine to 2-oxoglutarate resulting in the formation of pyruvate and glutamate as products. The pyruvate formed is reduced to lactate in the presence of NADH + H$^+$ by lactate dehydrogenase, added to the assay system. The reaction is followed spectrophotometrically by measuring the decrease in absorbance of NADH consumed in the reaction, at 340 nm. Stoichiometrically, the fall in absorbance is directly a measure of pyruvate formed during the course of the reaction.

$$\text{1. L-Alanine + 2-Oxoglutarate} \xrightleftharpoons{\text{ALT}} \text{Pyruvate + L-Glutamate}$$

LDH

2.   Pyruvate  Lactate

$$NADH + H^+ \quad NAD^+$$

Reagents: (i) L-Alanine-tris-HCl buffer, pH 7.4, 25 mM: Prepare 100mL of tris-HCl buffer (pH 7.4, 25 mM) containing 500 mM of L-alanine and 6.25μM concentration of EDTA (ii) Lactate dehydrogenase (source pig muscle, 2000 U/mL in glycerol or 500 U/mg protein; free from ALT contamination) : Dilute the enzyme to get 100 U/0.05 mL, in distilled water (iii) β- NADH + H$^+$ solution (sodium salt) 3.6 mM: Prepare freshly, 3 mg/mL solution of NADH + H$^+$ in alanine-tris buffer (pH 7.4, 25 mM) (iv) 2-oxoglutarate (α- Ketoglutarate) sodium salt: Prepare, 150 mM solution of 2-oxoglutarate in distilled water and adjust the pH to 7.4.

Procedure: Dispense 2 mL of L-Alanine-tris-HCl buffer into a clean test tube. Add sequentially, 0.05 mL of lactate dehydrogenase enzyme, 0.1 mL of NADH solution and 0.15 mL of test serum. Mix gently and incubate the contents for 10-15 min at 37°C in a constant temperature water bath. Transfer the contents to a 10 mm path length quartz cuvette (3 mL capacity) and immediately add 0.2 mL of 2-oxoglutarate solution and mix. Record the change in absorbance ($\Delta A/min$), using a UV-Vis spectrophotometer at λ 340 nm, at 37°C. Monitor the absorbance every 1 minute, over a period of 10 minutes. Calculate the $\Delta A/min$ between the time points of $3 - 9$ minutes. A lag time of 2-3 minutes may occur during initial stage of the enzymatic reaction.

Protocol*

Dispense successively	Volume (mL)
1. Assay medium containing, a) L-Alanine-tris-HCl Buffer (pH 7.4) b) Lactate dehydrogenase (LDH) c) NADH + H$^+$ solution	 2.0 0.05 0.10
2. Serum	0.15
Mix and pre-incubate at 37° C for 10 - 15 min	
3. 2-oxoglutrate (α- Ketoglutarate) solution	0.2

* Run the assay in duplicate.

Calculation

$$\text{ALT activity (UL}^{-1}) = \frac{(\Delta A_{340}/\text{min})}{6.22*} \times$$

$$\frac{2.5 \text{ mL (Total volume in the cuvette)}}{0.15 \text{ (Vol. of test serum used)}} \times 10^3$$

* *Molar absorption coefficient of NADH = 6.22 x 10³ mol⁻¹ cm⁻¹*

Experiment #11: **Assay of serum aspartate aminotransferase** (AST), earlier known as serum glutamate-oxaloacetate transminase (SGOT) (EC 2.6.1.1)[§]

Principle: The determination of serum aspartate aminotransferase (AST) activity is based on coupled enzyme assay. This enzyme catalyses the transfer of amino group from aspartate to 2-oxoglutarate resulting in the formation of oxaloacetate and glutamate as products. The oxaloacetate formed is reduced to malate in the presence of NADH + H⁺ by malate dehydrogenase, added to the assay system. The reaction is followed spectrophotometrically by measuring the decrease in absorbance of NADH consumed in the reaction, at 340 nm. Stoichiometrically, the fall in absorbance is directly a measure of oxaloacetate formed in the course of the reaction.

$$\text{AST}$$
1. L-Aspartate + 2-Oxoglutarate ⟷ Oxaloacetate + L Glutamate

$$\text{MDH}$$
2.   Oxaloacetate ⟷ Malate
               NADH + H⁺   NAD⁺

Reagents: (i) L-Aspartate-tris-HCl buffer, *p*H 7.6, 25 mM: Prepare 100mL of tris-HCl buffer (*p*H 7.6, 25 mM) containing 250 mM of L-aspartate and 6.25 µM concentration of EDTA (ii) Malate dehydrogenase (source pig heart muscle, 1000 U/mg protein in glycerol): Dilute the enzyme to get 100 U/0.05 mL, in distilled water (iii) β- NADH + H⁺ solution (sodium salt) 3.6 mM: Prepare freshly, 3 mg/mL solution of NADH + H in aspartate-tris buffer (*p*H 7.6, 25 mM) (iv) 2-oxoglutarate (α- Ketoglutarate) sodium salt: Prepare, 150 mM solution of 2-oxoglutarate in distilled water and adjust the *p*H to 7.6.

Procedure: Dispense 2 mL of L-aspartate-tris-HCl buffer into a clean test tube. Add sequentially, 0.05 mL of malate dehydrogenase enzyme, 0.1 mL of NADH solution and 0.15 mL of test serum. Mix gently and incubate the contents for 10 -15 min at 37°C in a constant temperature water bath. Transfer the contents to a 10 mm path length quartz cuvette (3 mL capacity) and immediately add 0.2 mL of 2-oxoglutarate solution and mix. Record the change in absorbance (Δ*A/min*), using a UV-Vis spectrophotometer at λ 340 nm, at 37°C. Monitor the absorbance every 1 minute, over a period of 10 minutes. Calculate the Δ*A/min* between the time points of 3 – 9 minutes. A lag time of 2-3 minutes may occur during initial stage of the enzymatic reaction.

Protocol *

Dispense successively	Volume (mL)
1. Assay medium containing,	
a) L-Aspartate-tris-HCl Buffer (pH 7.6)	2.0
b) Malate dehydrogenase (MDH)	0.05
c) NADH + H⁺ solution	0.10
2. Serum	0.15
Mix and pre-incubate at 37° C for 10 - 15 min	
3. 2-oxoglutarate (α- Ketoglutarate) solution	0.2

* Run the assay in duplicate.

Calculation

$$\text{AST activity (UL}^{-1}) = \frac{(\Delta A_{340}/min)}{6.22^*} \times \frac{2.5 \text{ mL (Total volume in the cuvette)}}{0.15 \text{ (Vol. of test serum used)}} \times 10^3$$

* *Molar absorption coefficient of NADH =* $6.22 \times 10^3 \, mol^{-1} \, cm^{-1}$

Clinical Significance of ALT and AST

Alanine aminotransferase is predominantly present in liver and to a lesser extent in kidneys, skeletal muscles and erythrocytes. The enzyme aspartate aminotransferase is widely distributed in all the body tissues, especially in heart and skeletal muscles and liver, where they occur in very high concentration. Elevated levels of both AST and ALT in serum are sensitive diagnostic markers for hepatocellular injury, however, serum ALT is a more specific indicator. In viral hepatitis, there is 10 -100 fold increase in transminase activity (both, ALT & AST). In alcoholic hepatitis serum AST activity tends to be higher than ALT. Rise in serum ALT and AST activities are noted in conditions such as skeletal muscle trauma, muscular dystrophy and myocardial infarction. In primary hepatocellular carcinoma, plasma AST activity is greater than ALT by 5-10 folds, due to hepatocellular damage. Moderate elevation in serum AST levels have been found in cases of gangrene, pancreatitis and hemolytic disease. Urinary excretion of aminotransferase is observed in patients with kidney lesions. Normal serum reference values of AST and ALT in adults are 6-18 and 3-26 UL^{-1}, respectively (enzyme assay at 30°C, without pyridoxal phosphate as a coenzyme).

Experiment #12: Assay of serum lactate dehydrogenase (LDH) (EC 1.1.1.27, L- Lactate NAD^+ - oxidoreductase) [§]

Principle: Pyruvic acid in the presence of $NADH + H^+$ is reduced to lactic acid by the catalytic action of the enzyme lactate dehydrogenase. The reaction is terminated by the addition of 2,4 - dinitrophenylhydrazine reagent, which condenses with the ketonic group of unreacted pyruvic acid, forming a hydrazone derivative. This hydrazone turns brown in alkaline medium, the colour of which can be colorimetrically measured at 510 nm.

Pyruvic acid   2, 4, - Dinitrophenyl hydrazine   2, 4- Dinitrophenylhydrazone of pyurvic acid

Reagents: (i) Phosphate buffer (*p*H 7.4, 100 mM): Prepare 100 mL of potassium-sodium phosphate buffer, *p*H 7.4, 100 mM (use anhydrous sodium phosphate (dibasic) salt and anhydrous potassium phosphate (monobasic) salt for preparing this buffer). (ii) Stock pyruvate (substrate) solution (37.5 mM): Dissolve, 415 mg of sodium pyruvate in 100 mL of phosphate buffer (aliquot and store the solution in a deep freezer) (iii) Buffered substrate solution: Dilute 1 mL of the stock substrate solution to 50 mL with phosphate buffer (to be prepared freshly). (iii) NADH + H$^+$ solution: Prepare 10 mg/mL solution in phosphate buffer (to be prepared freshly) (iv) 2,4 - dinitrophenylhydrazine reagent: Prepare 0.04% w/v solution by dissolving 40 mg of 2,4 - dinitrophenylhydrazine in 8.5 mL of conc. HCl and make up the volume to 100 mL with distilled water and store in a amber coloured reagent bottle (v) Sodium hydroxide solution (0.4 N).

Procedure

(i)   Take three clean test tubes and label them as Blank (B), Control (C) and Test (T). Dispense the assay reagents and test serum in the following sequence:

   Blank(B): Add, 1.2 mL of phosphate buffer and 1 mL of dinitrophenylhydrazine solution.

   Control(C): Add, 1.0 of buffered substrate, 0.2 mL of phosphate buffer and 1 mL of dinitrophenylhydrazine solution.

   Test (T): Dispense 1 mL of buffered substrate and 0.1 mL of test serum into to the tube and place the contents in a constant temperature water bath at 25°C, for 10-15 min. Initiate the enzymatic reaction by the addition of 0.1 mL of NADH solution. Incubate for exactly 15 min, at 25°C. Remove the tube from the water bath, and add 1 mL of dinitrophenylhydrazine solution and vortex.

Note: Serum LDH is sensitive to heat, hence the reaction is performed at 25°C.

Protocol:

Dispense successively (mL)	Blank (B)	Control (C)*	Test (T)
1. Buffered substrate	- -	1.0	1.0
2. Serum	- -	- -	0.1
3. NADH + H⁺ solution	- -	- -	0.1
4. 2,4 - dinitrophenylhydrazine reagent	1.0	1.0	1.0
5. Sodium hydroxide (0.4 N)	10.0	10.0	10.0
6. Phosphate buffer	1.2	0.2	- -

* Control tube contains 0.75 µmoles of pyruvate

(ii) Allow the test tubes (B, C and T) to stand at room temperature, for 15-20 min. Later add 10 mL of sodium hydroxide (0.4 N) solution to each tube and mix. Read the absorbance of the coloured solution after 10 min, at 510 nm.

Calculation: Control (C) tube contains 0.75 µmoles of pyruvate. The amount of pyruvate which has reacted is given by,

$$\frac{(C - T)}{(C - B)} \times 0.75 \ \mu moles$$

This is the effect of enzyme present in 0.1 mL of the serum, acting for 15 min. Therefore, pyruvate reacting per minute per liter of the test serum is given by,

$$\frac{(C - T)}{(C - B)} \times 0.75 \ \mu moles \ \frac{1}{15} \times \frac{1000}{0.1}$$

$$Serum \ LDH \ activity \ (UL^{-1}) = \frac{(C - T)}{(C - B)} \times 500$$

Clinical significance: Lactate dehydrogenase is a cytosolic enzyme with a relative molecular mass of 13.4 kDa and is comprised of four polypeptide chains of two types, namely M and H types. Five isoenzymes of LDH are known to exist (LD_1 to LD_5). This enzyme is predominantly present in heart, skeletal muscles, liver, kidneys and erythrocytes. Any damage to these tissues results in the release of enzyme into peripheral circulation, consequently a raise in serum LDH levels. Thus, this enzyme is an important diagnostic marker for the evaluation of myocardial infarction and muscle dysfunction.

Activity of LDH in normal serum is relatively low. An elevated enzyme level is observed in myocardial infarction within 6-12 h, reaches a maximum at 48-72 h, and remains elevated for 7-12 days. Serum LDH levels are also enhanced in muscular dystrophy and liver diseases such as hepatitis, liver cell necrosis, cirrhosis, liver tumors and obstructive jaundice. Abnormal levels of this enzyme are also associated with hematological (ex. Lymphomas, leukemia, pernicious and hemolytic anemia) and renal diseases like tubular cell necrosis, renal infarction and renal tumors. Artefactual elevated levels of serum LDH has been observed in hemolysed blood samples, possibly due to their release from erythrocytes. In chronic *glomerulonephritis, diabetic nephro-sclerosis* and kidney malignancy, LDH is excreted in urine. Reference LDH value in serum is in the range of 100 - 225 UL^{-1} (enzyme assay at 37°C).

Workout: Determine the activity of the above diagnostic enzymes in serum specimens of clinical origin.

§ (i) Wootton, I. D. P. Micro-analysis in Medical Biochemistry. 4th ed. J. A. Churchill Ltd. (1964).
(ii) Expert Panel on Enzymes (IFCC): Methods for the Measurement of Catalytic Concentrations of Enzymes. Clin. Chem. Acta. 70: F19-42, (1972).
(iii) Bergmeyer, H. U. (Editor-in-Chief), Methods of Enzymatic Analysis. Vol. III & IV. Verlag Chemie GmbH, Weinheim, Germany, (1983).

C. GASTRIC SECRETION

Experiment #13: Determination of gastric juice acidity

Principle: Gastric acidity is due to hydrochloric acid present in free as well as in combined form with mucin. Organic acids also contribute to total gastric acidity. This acidity can be determined titrimetrically with

standardized alkali using specific indicators.
(**Note**: Free acidity = acidity due to HCl; Total acidity = Acidity contributed by organic acids and HCl).

Reagents: (i) Standardised sodium hydroxide (0.1 N) solution (ii) Phenolphthalein indicator (1% w/v in ethanol) (iii) *Topfer's* reagent: Dimethylaminoazobenzene (0.5% w/v in 95% (v/v) ethanol).

Procedure

(i) *Total acidity*: Take 10 mL of gastric juice in a 100 mL conical flask and add 2 - 3 drops of phenolphthalein indicator and titrate the contents against 0.1 N NaOH to a pale pink end point. Record the titer value

(ii) *Free acidity due to HCl*: Separately, repeat the above procedure using 2-3 drops of *Topfer's* reagent as an indicator. A change from red to orange-yellowish colour indicates the end point. Record the titer value.

Calculation: Express the acidity value as $mEqL^{-1}$ of gastric juice, using the following relationship:

Total gastric acidity $\quad$ = $\quad$ Titer value (mL) x 100

Free acidity due to HCl $\quad$ = $\quad$ Titer value (mL) x 100

Note: one mEq of NaOH (40 mg) = 36.5 mg of HCl (1 mEq of HCl)

Clinical significance

Normal volume of gastric secretion for a 24 h period is in the range of 2 – 3 L. Clinically, an increase in gastric secretion of free HCl is known as *hyperchlorhydria*. This condition is observed in majority of the subjects suffering from duodenal ulcers. Measurement of gastric secretion is a useful diagnostic indicator of *achlorhydria*, a state characterized by an abnormal *p*H (6.0) of gastric secretion and absence of chloride ions. Majority of the gastric carcinoma patients suffer from *achlorhydria*. The normal range of free and total acid for 24h period is in the range of 18-25 and 28-48 $mEqL^{-1}$, respectively.

Workout: Determine the total gastric acidity value of a normal test specimen.

D. URINE ANALYSIS

Urine is an important excretory biological fluid that is formed by the kidneys. Presence of certain metabolites in urine reflects the physiological and metabolic status of the body. Further, the urine specimen can be easily collected for routine examination and thus is non-invasive in nature. Microscopic and biochemical investigation of urine is an important tool in the diagnosis and in obtaining information of various pathological conditions. Normally, complete urinary report includes, (i) physical and microscopic examination (ii) qualitative and quantitative analysis of urinary constituents.

Collection and preservation of urine: The urine sample used for clinical examination should be preferably collected over a period of 24 hours in a clean amber coloured bottle, containing suitable preservatives, such as four drops of 40% (v/v) formalin (formaldehyde solution) or 0.1 mg of thymol for 100 mL of urine. In addition, organic solvents such as toluene (methyl benzene), chloroform and petroleum benzene are added to layer the surface of the sample to avoid microbial contamination. Acids such as hydrochloric acid (10 mL / 24 hour urine sample) and glacial acetic acid (10 mL for 24h urine sample) are also used in preserving urine samples. The selection of a urinary preservative depends on the type of analysis. For example, thymol treated urine are suitable for the analysis of several urinary metabolites such as urea, creatinine, glucose, ketone bodies, calcium, potassium, sodium, amino acids etc., Formalin or chloroform treated urine sample are not preferable for glucose determination, as these preservatives also have the ability to reduce the oxidising reagents (*Benedict's* and *Fehling's* reagents). Urine specimens containing acids as preservative are suitable for the quantitation of urea, calcium, total nitrogen and ammonia.

(i) *Physical examination*

Colour and odour: Normal urine is colourless to straw coloured. The colour of the urine is due to the presence of urochrome, which is an oxidized form of urochromogen. Other pigments that contribute to the colour of urine are urobilin, uroerythrin, heamatoporphyrin etc. The normal colour of the urine is influenced by the volume of urinary output. Abnormal urinary colour is seen in various pathological conditions. Brown to black coloured is noted in clinical conditions such as *porphyrinuria, methaemoglobinuria, melanotic sarcoma* and in *alcaptonuria* - an inborn error of phenylalanine metabolism. A red to brown coloured urine is seen in conditions such as *haematuria* and *myoglobinuria*. A deep yellow or greenish-yellow coloured urine is observed in jaundice and in severe dehydration.

An aromatic odour is characteristic of freshly voided urine, due to the presence of certain volatile organic acids. Urine upon standing, develops an ammonical odour due to bacterial action on urinary urea. Similar ammonical odour is observed in subjects suffering from certain urinary tract infections. Infection due to Gram negative bacteria imparts an unpleasant odour to urine. In diabetics, excretion of ketone bodies (*ketonuria*) imparts a fruity odour to urine.

Volume: The daily output of urine in normal individuals ranges between 0.6 - 2.5 L /day. However, depending upon the physiological and pathological conditions, the 24 hour urinary volume may vary considerably, the mean output being 1.2 L/day. Various factors influence urinary output. For example high dietary intake of proteins has a diuretic effect leading to increased urinary volume, while high ambient temperature reduces output of urine. *Polyuria* - a clinical condition wherein there is an increased urinary output (> 3 L/day). This condition is observed in *Diabetes mellitus* and *insipidus*, chronic renal failure and ingestion of diuretic drugs. *Olgiuria* is a condition which is characterized by a decreased urinary out put (< 400 mL/day) and is seen in acute renal failure, severe dehydration and *oedema*. Presence of bilateral kidney stones, shock, and acute *nephritis* lead to cessation of urine excretion - a condition termed as *anuria*.

Turbidity: Freshly voided urine is clear and transparent. Flocculation of normal urine on standing is attributed to the presence of mucoproteins, nucleoproteins and epithelial cells originating from genito-urinary tract. Turbidity of urine is due to presence of pus cells, crystals and chyle that can be confirmed by microscopic examination. Precipitation of calcium phosphate, ammonium urate in alkaline urine and uric acid in acidic urine, also results in turbidity.

Specific gravity: The specific gravity of normal urine ranges from 1.003 - 1.030 and is measured by urinometer. Polyuria results in urine of low specific gravity. On the contrary, in diabetic polyuria, the specific gravity of urine is high due to presence of glucose and ketone bodies. Values greater than

1.030 are observed in clinical conditions such as adrenal insufficiency (*Addison's* disease), diabetes and nephrotic syndrome. Lowered specific gravity of urine is seen in *Diabetes insipidus* and chronic renal failure.

pH: Normal urine is acidic. The mean *p*H value of a 24 hour normal urine is 6.0, with a range of 4.5 - 8.5. The acidity of urine increases in fever, metabolic acidosis, respiratory acidosis, and after a protein rich meal. Alkalinity of the urine upon standing is due to the formation of ammonia from urea by the action of microorganisms and subsequent loss of CO_2. Excretion of alkaline urine is characteristic of urinary tract infections (UTI). Table 11.3 gives the general physical features of normal urine. Table 11.4 shows some important constituents of normal and abnormal urine.

Table: 11.3 General features of normal urine

Volume	0.6 - 2.5 L / Day or 24 h urine
Specific gravity	1.003 -1.030
*p*H	6.0 (mean value)
Total solids	180.0 - 220.0 g/L
Titrable acidity	25.0 - 70.0 mEq* of NaOH

* 1 mEq of NaOH = 40 mg of NaOH

(ii) *Microscopic examination*

Urine specimens can also be subjected to microscopic analysis as a preliminary screening procedure to evaluate any abnormality. For microscopic examination of urine, a sample of 10 mL urine is centrifuged at 3500 x g for 15 min. The supernatant is discarded and the sediment is resuspended in few drops of the supernatant urine. Transfer a small aliquot of the suspension onto a microscopic slide, covered with a glass cover slip. View the specimen under the light microscope, in both low and high magnification. Table 11.5 depicts the sedimented urinary constituents and its clinical relevance.

Clinical Biochemistry

Table 11.4 Normal and abnormal urinary constituents

Organic	24 hour urine / or Day	Inorganic	24 hour urine/ or Day
Allantoin	25.0 - 35.0 mg	Arsenic	5.0 - 50 µg
Amino nitrogen	0.14 - 1.5 g	Calcium	0.4 - 0.56 g
Catecholamines *	< 100 µg	Chloride	3.9 - 8.88 g
Cortisol	< 50 µg	Copper	3.0 - 35 µg
Creatine	Male: 0 - 40 mg Female: 0 - 80 mg	Fluoride	0.2 - 3.2 mg
Creatinine	1.0 - 2.0 g	Iodine	35.0 - 75.0 µg
Estrogen	Male: 15.0 - 40.0 µg Female: 15.0 - 80.0 µg	Iron	< 1.0 mg
Formiminoglutamic acid	35 mg	Magnesium	4 8..0 - 144.0 mg
Glucose	< 0.05 g	Phosphorus (inorganic)	0.7 - 1.5 g
Glutamine	44.0 - 152.0 mg	Potassium	1.2 - 4.8 g
Glycine	59.0 - 294.0 mg	Sodium	3.0 - 5.0 g
Glycolic acid	15.0 - 60.0 mg	Sulphates (total)	0.6 - 1.0 g
Hippuric acid	0.7 g	**Abnormal constituents of urine**	
17-hydroxycorticosteroids	Male: 3.0 - 10.0 mg Female: 2.0 - 8.0 mg	Abnormal constituent	Clinical condition
5-Hydroxyindoleacetic acid	2.0 - 7.0 mg	**Amino acids** Aminoaciduria	Familial disorders of amino--acid metabolism (Ex. Phenylketonuria, Cystinuria)
Hydroxyproline	0 - 1.3 mg		
17-ketosteroids	Male: 10.0 - 25.0 ng Female: 6.0 - 14.0 mg	**Bile pigments** Bilirubin	Jaundice
Lactic acid	0.5 - 1.98 mg	**Blood**	Haematuria
Niacin	2.4 - 6.4 mg	**Ketone bodies** Aceto acetic acid β-Hydroxybutyric acid	Starvation, Diabetes Ketonuria
Oxalic acid	20.0 - 60.0 mg		
Phenolic compounds	15.0 - 40.0 mg	**Porphyrins** Uro & Copro- porphyrins	Porphyrias
Phenylalanine	< 16.5 mg		
Protein (total)	< 50 mg	**Proteins** Albumin Haemoglobin Myoglobin Bence-Jones protein Glycoprotein (Mucus)	Albuminuria Haemoglobinuria Muscle injuries Multiple myeloma Urinary tract inflammation
Purine bases	< 4 0.0 mg		
Pyruvic acid	25.0 - 88.0 mg		
Urea	20.0 - 35.0 g		
Urea nitrogen	12.0 - 20.0 g	**Reducing sugars** Glucose Lactose Galactose Fructose Pentoses	Glycosuria, Diabetes Lactosuria Galactosuria Fructosuria Pentosuria
Uric acid	250.0 - 750.0 mg		
Vanilmandelic acid	1.4 - 6.5 mg		

Table 11.5 Urinary sedimer¹s and their clinical relevance

Urinary sediment	Clinical relevance
1. **Cells** (a) Erythrocytes (b) Leucocytes/pus cells (c) Small polygonal epithelial cells	Glomerulonephritis Urinary tract infection Renal dysfunction
2. **Casts** (These are cylindrical masses formed in the renal tubules by the agglutinated proteins/ mucoproteins (*Tamm-Horsfall* protein)/ cells). Types of casts (a) *Hyline casts*: Formed by albumin. (b) *Epithelial casts*: Formed from desquamation of tubular epithelium in large amounts. (c) *Cellular casts*: Formed by RBC & WBC. (d) *Fatty casts*: Derived from interaction of epithelial casts with fat globules along with certain granular material. (e) *Granular & Waxy casts*: Derived from degradation of cellular casts during their movement in the renal tubules give rise to granular and later to waxy casts.	 *Albuminuria* Renal tubular damage Acute *glomerulonephritis* & *Pyelonephritis* Degenerative disease of tubular epithelium. Tubular dysfunction
3. **Crystals** Deposits of salt crystals of phosphates and oxalates in acidic and alkaline urine. Urates.	*Nephrolithiasis* (renal calculi) Gout, Gouty arthritis

(iii) *Qualitative tests for abnormal urinary constituents*:

Tests for Proteins

(a) *Coagulation test*

Principle: Proteins upon heating undergo coagulation and precipitation. This phenomena is due to denaturation of proteins and the method is useful in identifying, urinary proteins such as albumin & globulins under pathological conditions.

Procedure: Take 5 mL of the urine specimen into a test tube and add 4 to 5 drops of glacial acetic acid and mix. Using a test tube holder, heat the contents over a Bunsen flame to boiling.

Observation: Appearance of turbidity or a flocculant precipitate is indicative of protein.

Note: Acidification of urine is done to avoid interference of urinary phosphates which precipitate upon heating.

(b) *Sulfosalicylic acid test*:

Principle: Proteins get precipitated upon addition of chemical agents such as sulphosalicylic acid, phosphotungstic acid etc., Hence, this test can be used in the detection of proteins in pathological urine.

Reagent: Sulphosalicylic acid solution (25% w/v).

Procedure: Transfer 10 mL of the urine specimen into a test tube. Add 0.1 mL of sulphosaliylic acid reagent and mix.

Observation: Appearance of turbidity or white precipitate suggests the presence of proteins.

(c) *Test for Bence-Jones protein*:

Principle: These abnormal proteins are excreted in urine of patients suffering from multiple myeloma. *Bence-Jones* protein represent the light chains of immunoglobulins. They are detected by their characteristic behaviour upon heating. The proteins undergo thermal denaturation at 40-60°C and precipitate, which redissolves upon increasing the temperature (boiling). The proteins reprecipitate upon cooling.

Procedure: Take 5 mL of the urine specimen into a test tube and heat the contents in a water bath, maintained at 60°C. Note the change in the appearance. Heat the contents further to a boiling temperature and observe.

Observation: Appearance of a precipitate followed by its disappearance upon increase in temperature confirms the presence of *Bence-Jones* protein.

Test for sugars

Benedict's test

Principle: Presence of urinary reducing sugars, such as glucose is detected by *Benedict's* test. In this test the cupric copper of the oxidising reagent gets reduced to cuprous oxide by the sugar. During the reaction, the sugar gets oxidised to sugar acid (gluconic acid).

Reagent: (i) *See*, *6. **Qualitative analysis**,* for preparation of the *Benedict's* reagent (ii) NaOH solution (1 N).

Procedure: Take 0.5 mL of the urine sample in a test tube. Make the content alkaline by adding a drop of 1 N NaOH solution. Add 5 mL of *Benedict's* reagent and heat the contents in a boiling water bath for 3 - 5 min.

Observation: A clear blue solution indicates absence of sugar, while a precipitate with varying colour (green/orange/red) depending upon the amount of sugar present, confirms the presence of reducing sugar (glucose).

Note: In addition to glucose, other reducing sugars*, such as xylose, fructose, galactose and lactose respond to this test. Abnormal urinary metabolites such as homogentisic acid (excreted in aromatic amino acid metabolism - Alcaptonuria) also reduce *Benedict's* reagent.
*Perform specific tests such as, Bial's-orcinol , Seliwanoff's, osazone and mucic acid tests for the identification other reducing sugars (see, **6. Qualitative analysis for details**).*

Test for Haemoglobin

Benzidine test

Principle: Presence of haemoglobin in urine (*haematuria*) is detected by benzidine reaction. In this reaction, haemoglobin catalyses the generation of nascent oxygen from hydrogen peroxide, which in turn oxidises the colourless benzidine (p, p -diamino-biphenyl) to a blue coloured compound.

Reagent: (i) Benzidine reagent: Dissolve 1g of benzidine dihydrochloride in distilled water and make up the volume to 100 mL(1% w/v) (ii) Hydrogen peroxide solution 3% (v/v) [make a 1:10 dilution of commercial H_2O_2 (30% v/v) with distilled water]. Both the reagents to be freshly prepared. (iii) Sodium acetate solution (1 % w/v).

Caution: Handle benzidine (toxic chemical) and H_2O_2 (oxidising agent) with utmost care, by wearing disposable gloves).

Procedure: Transfer 2 mL of the urine specimen into a clean test tube. Add 1 mL of benzidine reagent, 1 mL of 3% H_2O_2 solution, followed by addition of 1mL of 1% sodium acetate solution and mix the contents.

Observation: Development of blue or green colour after few minutes indicates the presence of haemoglobin.

Test for ketone bodies

Ketone bodies include acetoacetic acid, β-hydroxybutyric acid and acetone, which are formed during the course of lipid metabolism. The presence of urinary ketone bodies is detected by colour tests, such as *Rothera's* and *Gerhardt's* tests.

(a) *Rothera's test*

Principle: Ketone bodies react with sodium nitroprusside under alkaline condition to yield a purple colour.

Reagents: (i) Sodium nitroprusside (10 % w/v) solution (freshly prepared) (ii) Chilled liquor ammonia (iii) Ammonium sulphate (solid).

Procedure: Saturate 5 mL of the urine sample with solid ammonium sulphate in a test tube. Add 2 - 3 drops of sodium nitroprusside reagent followed by addition of few drops of chilled liquor ammonia and mix.

Observation: Appearance of a purple colour confirms the presence of ketone bodies.

(b) *Gerhardt's test*

Principle: Acetoacetic acid - a keto acid exhibits keto - enol tautomerism. The enolic form of this acid reacts with ferric chloride to yield a purple colour, which forms the basis for this test.

Note: Phenolic drugs such as salicylates, phenacetin, antipyrine etc., respond to this test.

creatinine ranges from 1-2 g. Creatinine clearance is a good measure of glomerular filtration rate. Creatinine clearance by kidneys is reduced in all conditions where, glomerular filtration rate is impaired, as observed in dehydration, acute & chronic renal failure and in obstructive uropathies. In chronic renal failure, the creatinine value of blood can exceed 30 mg/dL. Plasma or serum creatinine concentration is a reliable indicator of renal function. The excretion of creatinine in urine is constant and is related to muscle mass.

Workout: Determine the urinary creatinine content of 8, 16, and 24 hour human urine specimen.

[§] Bonsnes, R. W., & Taussky, H. H. J. Biol. Chem. 158: 581, (1945).

Experiment #15: Determination of urinary titratable acidity

Principle: The acidity of urine is contributed mainly by phosphate ions that can be titrated by using standardized alkali and expressed in terms of alkali consumed. Urinary acidity is also due to the presence of certain organic acids. However, their contribution to titratable acidity is negligible, as they exist as salts.

Reagents: (i) Standardized 0.1 N NaOH (ii) Phenolphthalein indicator (1% w/v) solution. (iii) Powdered potassium oxalate salt.

Procedure

(i)   Transfer 25 mL of fresh urine specimen into a clean 250 mL Erlenmeyer flask. Add 5 g of potassium oxalate salt to the contents directly and mix. To avoid the formation calcium phosphate at point of neutralization, urinary calcium is precipitated by addition of potassium oxalate to an insoluble calcium oxalate. This step prevents interference of urinary calcium ions.

(ii)   Add five drops of phenolphthalein indicator and titrate the contents against standardized 0.1 N NaOH to a pale pink end point. Note the volume of NaOH consumed.
(One mEq of NaOH = 40 mg of sodium hydroxide).

Calculation

$$\text{Urinary acidity } (Z) = \frac{XY}{25 \text{ (Volume of urine taken - mL)}}$$

X = Titer value (mL); Y = 24 hours urinary volume (mL); Z = Urinary acidity (mEq)

Clinical significance

The titratable acidity of normal urine ranges from 25.0 - 70.0 mEq of NaOH. This value is influenced by the type of diet. The acidity of urine is enhanced in clinical conditions such as renal tubular acidosis and severe diabetes.

Workout: Express the urinary acidity in terms mEq NaOH for 24 h urine sample.

Note: One mEq of NaOH = 40 mg of sodium hydroxide.

(Method for analysis of urinary glucose, urea, bilirubin, calcium - *See 7.* *Quantitative Analysis and this chapter*)

Additional Reading

1.   Varley, H. Gowenlock, A. H. & Bell, M. Practical Clinical Biochemistry. 5th ed. William Heinemann Medical Books Ltd. London, (1980).

2.   Robert, B. (editor-in-chief), The Merck Manual of *Diagnosis and Therapy*. Merck Sharp & Dohme Research Laboratories, Division of Merck & Co. Inc. Rahway, NJ, USA, (1982).

3.   Hector, T., & James, A. M. Assessment and Control of Biochemical Methods. John Wiley & Sons, Chichester, UK, (1986).

4.   Hawcroft, D. Diagnostic Enzymology. (ed), James, A. M.. John Wiley & Sons, Chichester, UK, (1987).

5.   Taylor, E., (ed), Clinical Chemistry, Wiley-Interscience, New York, USA, (1989).

6.	Kaplan, L., & Pesce, A. Clinical Chemistry: Theory, Analysis and Correlations. 3rd ed. C. V. Mosby, St. Louis. USA, (1996).

7.	*Tietz* Fundamentals of Clinical Chemistry. 5th ed. Burtis, C.A., & Ashwood, E. R., (eds). W. B. Saunders Company/ Harcourt (India) (P) Ltd., New Delhi, (2001).

Lab Notes

Lab Notes

12

Immunochemical Methods

Immune system is a specialised function expressed in vertebrates, which enables them to distinguish between *self* and *non-self*. This recognition system in higher animals provides protection against a wide variety of infective agents that cause diseases. Immune response is mediated through (i) *humoral* and (ii) *cell-mediated* pathway. The body cells that are involved in immune response include, B-lymphocytes (bone marrow derived cells), T-lymphocytes (thymus derived cells) and antigen-presenting cells. Cellular co-operation of B- and T-cells, more specifically *T-helper cells* enhances the production of immunoglobulin (**Ig**), also commonly known as antibody (**Ab**). Antibodies exhibit high binding affinities to specific surface chemical motifs (*epitopes*) of the immunogen and this forms the basis for detection of antigen in biological fluids.

An extraordinary selective and versatile reagent is provided by nature in the form of an immunoglobulin. Antibodies with high specificity can be synthesized by an organism in reasonable quantity within few weeks of injection of an antigen (**Ag**). Immunoglobulins are glycoproteins, synthesized by plasma cells (differentiated B-cells). They are high molecular weight plasma proteins present in blood and mediate the immune response in vertebrates. Structurally, immunoglobulin molecule is a *heterodimer*, consisting of two identical light (**L**) polypeptide chains (MW 23 kDa) and two heavy (**H**) polypeptide chains (MW 50-75 kDa) held together (L_2H_2) by disulfide bridges. There are five classes of immunoglobulins [denoted by Greek letters - _mu_ (μ), *gamma* (γ), _alpha_ (α), _delta_ (δ) and _epsilon_ (ε)], **IgM, IgG, IgA, IgD** and **IgE**, based on the type of the heavy chain they carry. All immunoglobulin molecules contain a minimum of two **L**-chains and two **H**-chains. Immunoglobulins can be produced as components of *polyclonal* hyper-immune serum or as *monoclonal* antibodies. The former involves immunizing and subsequent bleeding of appropriate animal model (ex. rabbit, guinea pigs, sheep, goat etc.). The later is secreted by immortalized clone of plasma cell, which is achieved by laboratory manipulation.

The interaction of an antigen and immunoglobulin are highly specific and this knowledge of immunological reaction has been successfully used in developing immunoanalytical methods, such as gel-diffusion techniques, immunoelectrophoresis, immunoassays (radioimmunoassay, enzyme immunoassay) and immunoblotting techniques.

The interaction of Ag-Ab is dependent on the non-covalent forces that stabilize the interactions. These binding forces include, *van der Waals*, *electrostatic, hydrogen bonding* and *hydrophobic* interactions. Thus, these complexes of antigen-antibody can be readily dissociated at low or high *p*H or by high ionic strength. Antibody can interact with (i) soluble antigen or (ii) particulate antigen. Its interaction with soluble antigen results in precipitation reaction, while with particulate antigen leads to agglutination or clumping.

In this chapter few experiments are designed to visualize the antigen-antibody interaction, their isolation and application as a bioanalytical tool.

Experiment # 1: Demonstration of antigen-antibody interaction: *Ouchterlony* technique [§]

The immunological relationship between an antigen and its specific antibody can be assessed by setting up the precipitation reaction in agar or agarose gels. This technique is alternatively known as immunodiffusion method and is a useful qualitative test. The method finds application in determination of antibody titer of hyper-immune serum and in establishing the cross-reactivity of the antibodies.

Principle: The antigen and the specific antiserum diffuse out towards each other in the gel and at the point they meet, they interact with each other, forming immune-complex by cross linking and precipitating at the point of equivalence (*lattice formation*), thus resulting in the formation of a spur or a precipitin line (12.1 & 12.2). If the point of equivalence is not reached or if either of them is in excess, then the immune complex solubilizes.

Biochemicals: (i) Antigen - Rabbit IgG or Bovine serum albumin (ii) Antiserum - anti-rabbit IgG raised in goat or anti-bovine serum albumin raised in rabbit.

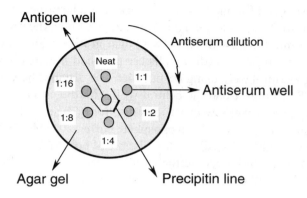

Fig. 12.1 Interaction of antigen-antibody in agar gel

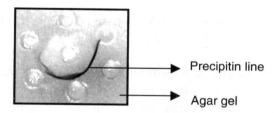

Precipitin line

Agar gel

Fig. 12.2 Antigen-antibody precipitin line stained with Coomassie brilliant blue R stain

Plastic ware: Plastic Petri-plates - 50 mm / 80 mm dia. (Polycarbonate), disposable polypropylene Eppendorf tubes (1.5 mL cap.) and a plastic bread box.

Equipment: Incubator, hot plate, autoclave, auto-dispenser (1-40 μL volume) and gel puncher (2 mm dia).

Chemicals/Reagents: (i) Agar (Bacto-grade) (ii) Phosphate buffered saline (PBS) 20 mM, pH 7.2, containing 0.05% sodium azide as an anti-microbial agent. Prepare 250 mL of PBS buffer.

Procedure: Pre-coat a disposable Petri-plate (2 mm thick) with 1% (w/v) agar in distilled water. For making the gel solution, the agar is added to water and steamed for one hour in an autoclave. The pre-coated Petri-plate is dried overnight, so as to get a thin film of agar. Later, the pre-coated plate is coated (2 mm thick) with 1.2% (w/v) agar in PBS buffer and allowed to solidify. Two millimeters (dia) wells are punched in a hexagonal pattern, containing a central well, using gel puncher. The distance between the wells should be 8 -10 mm. Load the central well with the antigen (20 – 25 μL) and the peripheral wells with various dilutions (neat, 1:1, 1:2, 1:4, 1:8, 1:16) of antiserum (20 -25 μL) in phosphate buffer, 20 mM, *p*H 7.2. Incubate the plate overnight (14 - 16h) in a moist bread box, at 37°C. At the end of the incubation period, observe the plate for precipitin line formation and record the titer of the antiserum tested (Fig. 12.1 & 12.2).

Workout: Determine the titer of specimen antisera.

§ Ouchterlony, Ö., Acta Pathol. Microbiol. Scan. 32: 231, (1953).

Experiment # 2: Demonstration of direct agglutination reaction: Determination of human blood group antigens *§*

Principle: Antibodies interact with multivalent particulate antigen, resulting in cross-linking of particulate antigen. This cross-linking of particulate antigen and antibody eventually leads to clumping or *agglutination*. Human erythrocytes possess surface glycoprotein antigen(s) A or B and AB, that agglutinate in the presence of specific anti-A or anti-B antiserum, indicating the presence of the respective antigen. Negative agglutination reaction indicates the absence of the antigens. This diagnostic test forms the basis of ABO blood grouping in humans.

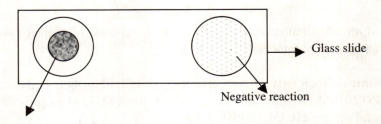

Positive agglutination reaction (clumping of erythrocytes in presence of antiserum)

Fig. 12.3   Agglutination reaction

Reagents: (i) Antibody (IgM-monoclonal, ex. murine) against human blood group antigen A & B (ii) Surgical spirit and sterile cotton.

Procedure

(i)   Place a clean glass slide on white ceramic tile. Mark out two areas on the slide and label them as 'A' and 'B'. Clean the tip of the left hand-ring finger with surgical spirit. Using a sterile lancet or a 22G sterile needle prick the tip of the left hand-ring finger and transfer two drops of blood separately on the marked areas. Stop the bleeding at the site of the needle prick by pressing a cotton swab dipped in surgical spirit.

(ii) Immediately add a drop of anti-A and anti-B antibodies separately, on to the blood drops and mix with a plastic tooth pick. After 2-3 min observe the agglutination reaction to identify the blood group (Fig. 12.3).

      Caution: (i) Do not reuse the used lancet or the 22G needle (ii) Destroy the needle and lancet immediately after use.

(iii) Interpretation of test results: (a) Agglutination of erythrocytes indicates a positive result (b) Absence of agglutination of erythrocytes indicates a negative result.

Major Human blood groups and their phenotype frequencies (%)

Type of blood group antigen	Antigen carried by erythrocyte	Circulating antibodies in blood	Reaction with anti-A antibody	Reaction with anti-B antibody	Phenotype Frequencies (%)
'O'	None	Anti- A & Anti-B	Negative	Negative	47
'A'	'A'	Anti-B	Positive	Negative	42
'B'	'B'	Anti-A	Negative	Positive	8
'AB'	'AB'	None	Positive	Positive	3

Note: Separately, by following the above procedure, presence of (Rh^+) or absence (Rh^-) of *Rhesus* (*Rh*) antigen (also called antigen-D) on human erythrocyte membrane can also be tested in the blood sample, using anti-D murine antibodies. (% frequencies - Rh^+ = 85; Rh^- = 15).

Workout: Determine the frequency distribution of blood group antigen among the fellow students.

[8](i) Mueller-Eckhardt, C. Blood Group Serology: *Laboratory Notes for Medical Diagnostics*. Behringwerke AG, Medizinische, Information / Eukerdruck KG, Marburg/ Lahn, Germany, (1975).

(ii) Davey, B. Immunology. Guidance Notes for Advanced Biology, No. 5. The Biochemical Society Publication, (BASC), Portland Press, London. (1994).

Experiment # 3. Demonstration of immuno-flourescence technique

Principle: Immobilized antigen spot on the nitrocellulose disc is visualized by fluorescent dye labelled antibody reagent, after blocking, under long wave UV light.

Biochemicals: Rabbit IgG (immunogen), bovine serum albumin, Fluorescein isothiocyanate labelled-goat antibody (IgG) raised against rabbit IgG (labelled antibody conjugate), Tween-20, and Tris.

Plastic ware/ matrix: Small glass or polypropylene Petri-plates (25 mm dia), auto-pipette tips. Nitrocellulose discs (12 mm dia; used for protein blotting). These discs can be prepared by a mechanically punching a rectangular strip of nitrocellulose.

Equipment: Laboratory incubator, refrigerator, auto-pipettes (variable volume, 20-200 & 200 -1000 μL), Long-wave UV (>380 nm) cabinet and table top platform shaker.

Reagents : (i) **Coating buffer** (50 mL): Carbonate-bicarbonate buffer, 50 mM, pH 9.6 (ii) **Incubation buffer** (100 mL): Tris-HCl buffer, 20 mM pH 7.4, containing 0.85% (w/v) NaCl (iii) **Washing buffer** (500 mL) : Tris-HCl buffer (20 mM, pH 7.4) containing 0.05%(v/v) Tween-20 and 0.05%(w/v) sodium azide (iv) **Blocking buffer** (50 mL): Tris-HCl buffer (20 mM, pH 7.4) containing 1 % (w/v) bovine serum albumin and 0.05%(w/v) sodium azide.

Procedure

(i)   Prepare 1 mg mL^{-1} concentration of rabbit IgG in carbonate-bicarbonate buffer (50 mM, pH 9.6). Dilute one volume of rabbit IgG with one volume of carbonate-bicarbonate buffer to give a 50% dilution. Load individually, 20 μL spots ($\geq$ 4 mm dia) of neat and diluted antigen on to the nitrocellulose discs and allow it to dry at room temperature. Handle the disc using a plastic forceps.

(ii)   Incubate the discs, in blocking buffer, taken in a Petri-plate, with constant shaking at 37°C for 3h, to block the non-specific binding of labelled antibody. Simultaneously, incubate a disc without antigen, which serves as control or blank. Remove the blocking buffer and wash the discs thrice (3X) with 10 mL washing buffer, so as to remove the unbound proteins.

(iii)   Incubate the discs separately in 2 – 3 mL incubation buffer containing 1:500 diluted fluorescein isothiocyanate labelled-goat antibody (IgG) raised against rabbit IgG, at 4°C, overnight (16 – 18h) in a refrigerator or in a cold room with constant shaking. After overnight incubation, remove the incubation buffer and wash the discs repeatedly (5X) with 10 mL of washing buffer and later air dry the discs. View the discs under long wave UV light, in a UV cabinet to visualize fluorescence in antigen coated discs.

Experiment # 4: Purification of bovine serum immunoglobulin G (IgG) fraction by ammonium sulphate precipitation (*micro-method*) [§]

Principle: Proteins in solution form hydrogen bonds with the surrounding aqueous *milieu*, through their exposed polar/ionic functional groups. When high concentration of small and highly charged ions such as ammonium and sulphate are added to protein solutions, they compete with the protein for binding to water molecules. This interaction decreases the solubility of the proteins and results in their precipitation. Precipitation of bovine IgG from bovine serum can be achieved by saturating the solution to 50% with saturated ammonium sulphate. The purity of the IgG isolated by this method is 90-95%.

Biochemicals: Bovine serum- 5 mL (Isolated serum from the blood, procured from a local *abattoir*).

Plastic ware: Graduated, Eppendorf tubes (2 mL capacity) or polypropylene centrifuge tubes (5 mL capacity).

Equipment:: pH meter, Cyclo-mixer, Table top centrifuge.

Reagents: (i) Saturated ammonium sulphate solution (dissolve 40 g of salt in 50 mL of warm distilled water and allow the salt solution to reach the room temperature. Filter and adjust the pH to 7.0 (using dilute ammonium hydroxide); store the reagent in a plastic container (ii) Phosphate buffer-100mL (50 mM, pH 8.0).

Procedure: To 1 mL of serum taken in an Eppendorf tube, add slowly, 1mL of saturated ammonium sulphate reagent, while vortexing the tube on a cyclo-mixer. After 10 min. centrifuge the contents for 10 min at 5000 rpm. Discard the supernatant and resuspend the precipitated protein in 1 mL of phosphate buffer (50 mM, pH 8.0). Reprecipitate the protein by adding slowly, 1mL of a saturated ammonium sulphate reagent, while vortexing. After 10 min, centrifuge the contents at 5000 rpm for 30 min. Discard the supernatant and redissolve the precipitated IgG fraction to a final volume of 1 mL in phosphate buffer. Dialyse the IgG fraction overnight in a beaker containing 2 L phosphate buffer (50 mM, pH 8.0, containing 0.05%(w/v) sodium azide) to remove ammonium sulphate. Alternatively, a desalting Sephadex G-10 column can be used to remove ammonium sulphate (*see, 8. Biochemical Separation Techniques*).

Workout: Estimate the total protein content in the serum and in the purified IgG fraction. Calculate and express the IgG content as % of the total serum proteins estimated.

Note: Ammonium sulphate interferes in protein estimation by Lowry method.

[§] Javois, L. C. (ed), Methods in Molecular Biology: Immunochemical Methods and Protocols. Vol. 24. Humana Press, Inc. Totowa, NJ, USA, (1994).

Experiment # 5: Rocket immunoelectrophoresis [s]

This technique is based on the combination of electrophoresis and immuno-precipitation of antigen-antibody complex in gels. This immuno-analytical method is both, qualitative and quantitative.

Principle: Rocket immunoelectrophoresis involves movement of immunogen into a gel containing dispersed antibodies that do not migrate under the applied current. The pH of the gel is the isoelectric pH of the immunoglobulin, hence the antibodies tend to be immobilized in the gel. The pattern of immunogen-antibody precipitation resembles a rocket, since precipitation occurs all along the moving boundary of the immunogen (Fig. 12.4). The height of the rocket, for a given antisera is linearly proportional to the concentration of the immunogen. The concentration of the test immunogen can be determined by interpolation from the reference standards (Fig. 12.5).

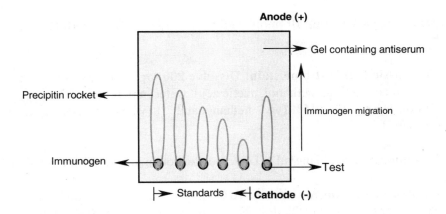

Fig. 12.4 Rocket immunoelectrophoresis

Biochemicals/chemicals: Bovine serum albumin (BSA), Antiserum to BSA-raised in goat or sheep, Agarose, Agar, sodium diethyl barbiturate or Tris, Coomassie Brilliant blue R-250, bromophenol blue, acetic acid, methanol, sodium chloride, and sodium azide.

Glassware: Glass slides (60 x 60 mm; thickness 2 mm), pipettes, conical flasks, beakers, measuring cylinders.

Plasticware: Flat bottomed plastic container and tray, auto-pipette tips, Eppendorf tubes (1.5 mL).

Equipment: Power pack, immuno-electrophoretic apparatus (horizontal type), auto-pipettes, gel puncher (dia. 2 mm). Laboratory oven, hot plate, hot water bath, level line, autoclave and refrigerator.

Reagents: (i) Barbital Buffer (50 mM, *p*H 8.6): Dissolve, 10.3 g of sodium diethyl barbiturate and 0.2 g sodium azide in 500 mL of distilled water. Adjust the *p*H to 8.6 with 6 N hydrochloric acid and make up the volume to 1 L (ii) Agarose (2.4% w/v): Disperse 2.4 g of agarose in barbital buffer and heat the solution in a boiling water bath until a clear solution is obtained (iii) Agar(1% w/v): Disperse 1g of agar in distilled water and steam the solution in an autoclave for 30 min. or (iv) Tris-glycine buffer (containing, 40 mM tris and 10 mM glycine), *p*H 8.6.

Tracker dye solution: Bromophenol blue (0.02%, w/v) in barbital or tris buffer

Coomassie brilliant blue stain: Dissolve 200 mg of the dye in 100 mL solution containing, methanol, acetic acid, water in the ratio of 5:1:5. *Destaining solution*: 15% (v/v) methanol and 7% (v/v) acetic acid in water (prepare 1 L).

Immunogen: BSA (6 mg/mL)**: Dissolve the protein in barbital or tris buffer.

Agarose containing antiserum: To 0.5 mL of neat antisera, add 9.5 mL of barbital or tris buffer (1:20 dilution of antisera) in a clean 20 mL glass test tube and mix. To 10 mL of 2.4% agarose solution kept at 56°C in a hot water bath, add 10 mL of 1: 20 diluted antiserum. The final dilution of antiserum in the agarose gel is 1:40, which is used for coating the glass slide. (**Note**: The agarose containing antisera should be used for coating within 15 - 20 min after preparation).

Procedure

(i)   Coat the glass slide with 1% agar solution by placing it on a horizontal surface (use level line). Allow it to solidify at room temperature. Later, dry the slide overnight in an oven maintained at 70°C.

(ii)   To the pre-warmed glass slide (50 - 60°C), coat agarose gel containing antisera, kept at 56°C, immediately and allow it to solidify. Place the glass slide in a refrigerator for 5 - 10 min and later punch six to eight wells of 2 mm dia. at a distance of 8 mm from the bottom of the slide. The wells should be punched in a straight line.

(iii)   Dispense 20-25 μL of the diluted reference standard immunogen (in the range of 0.5 - 5 mg BSA/mL), along with 5 μL of the tracker dye into the respective wells. Similarly, the unknown test samples can also be run along with reference standards.

(iv)   Place the slide horizontally in the electrophoretic apparatus in such a way that the wells are positioned close to the cathode (negative terminal). This step should be performed immediately, in order to avoid any radial diffusion of the immunogen present in the wells.

(v)   The electrophoretic tanks are filled with equal volumes of barbital or tris buffer. Filter paper (Whatman No.1) wicks are placed on both the ends of the slide. Later, the free ends of the filter paper wick are dipped in the electrophoretic tanks. Wet the filter paper with the buffer and connect the leads to the power pack, after closing the lid of the electrophoretic apparatus. Turn on the power pack and apply constant voltage of 3-4 V/cm for 4 to 5 hours.

(vi)   After the completion of the electrophoretic run, turn off the power supply and remove the slide and observe the rocket shaped precipitin patterns. Measure and record the height of the rocket pattern / distance of migration in millimetres. Construct a standard plot with height of the rocket *vs* dilution or concentration of the immunogen.

(vii)   Alternatively, the rocket precipitin pattern can be stained with coomassie blue dye for visualization. The developed slide is immersed and washed successively with 0.3 M and 0.15 M sodium chloride solution, twice in a plastic container, with gentle shaking, for 1- 2 hours. After washing, the wells of the slide are filled with distilled water and later a wet Whatman No.1 filter paper (60 x 60 mm) is covered on the slide and dried at 60°C, overnight in an oven. The dried slide is then stained by immersing it in coomassie blue solution present in a flat plastic tray, for 10 - 20 min. Later, the excess stain is removed by destaining solution (over a period of 3 to 5 hours, with 3 to 4 changes of destaining solution). Measure and record the height of the stained rocket pattern (blue colour) / distance of migration in millimetres. Construct a standard plot with height of the rocket *vs* dilution or concentration of the immunogen (Fig. 12.5).

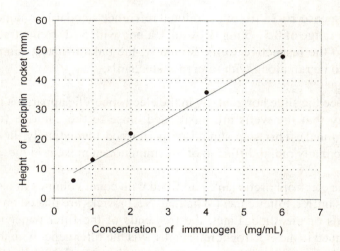

Fig. 12.5 Linear relation between immunogen concentration & height of
the precipitin rocket

§ Laurel, C. B. Anal. Biochem. 15: 45, (1966).

Experiment # 6: Enzyme linked immunosorbant assay (ELISA): Antibody capture assay §

Antibodies that exhibit high binding affinities to different surface chemical motifs of the immunogen are made use in the development of analytical immunoassays, such as RIA and ELISA. The basic concept involved in an immunoassay is the interaction of an immunogen with its specific antibody. In enzyme linked immunosorbant assay (ELISA), soluble immunogen is linked to an antibody (which is bound to a solid phase like microtiter plate, plastic beads or strips) or soluble antibodies to immobilized immunogen, in a manner that allows both immunological and enzymatic activity to be retained. The free immunogen (analyte) and the enzyme labelled immunogen compete for the fixed and limited number of specific binding sites on the immobilized antibody. Alternatively, the enzyme labelled antibodies can be made to compete with the immobilized immunogen and free immunogen (analyte). The net absorbance due to the immunogen or the antibody linked enzyme activity is correlated to the concentration of the analyte.

(i)

$$\text{Ab} + \text{I} + \text{I}^{enzyme} \leftrightarrow \text{Ab.I} + \text{Ab.I}^{enzyme}$$

(bound)   (free)   (free)   (bound)   (bound)

(ii)

$$\text{Ab}^{enzyme} + \text{I} + \text{I} \leftrightarrow \text{Ab}^{enzyme}.\text{I} + \text{Ab}^{enzyme}$$

(free)   (bound)   (free)   (bound)   (free)

[Ab - Antibody; I - Immunogen (analyte); I ^{enzyme} - Enzyme labelled immunogen; Ab^{enzyme} - Enzyme labelled antibody]

Antibody capture assay

Antibody capture assay is an immunoanalytical method, based on the principle of enzyme immunoassay. This technique is useful in detecting and quantitating the circulating levels of immunoglobulins, screening and detection of bacterial and viral antigens in blood plasma or serum. The method is simple, sensitive and specific with detection of analyte in the range of pg to ng.

Principle: Various concentrations of the immunogen are immobilized onto a polystyrene micro-titer plate along with the test immunogen. The bound immunogen is detected by the enzyme labelled-specific antibody to the immunogen. The enzyme activity due to the immunogen bound enzyme labelled antibody is directly proportional to the concentration of the immobilized immunogen.

Biochemicals: Rabbit IgG, bovine serum albumin, alkaline phosphatase labelled-goat antibody (IgG) raised against rabbit IgG, Tween-20, diethanolamine, *p*- nitrophenyl phosphate.

Plasticware: Polystyrene ELISA/microtiter plate (12 x 8 wells, flat bottomed), microfuge/ Eppendorf tubes (1.5 mL), auto-pipette tips.

Equipment: Laboratory incubator, auto-pipettes (variable volume, 40-1000 µL), spectrophotometer.

Reagents

(i)   **Coating buffer** (50 mL): Carbonate-bicarbonate buffer, 100 mM, *p*H 9.6.

(ii)   **Blocking buffer** (25 mL) : Phosphate buffered saline (PBS) 10 mM, *p*H 7.2, containing 0.2 % (w/v) bovine serum albumin and 0.05% (w/v) sodium azide.

(iii) **Washing buffer** (1L) : Phosphate buffered saline-Tween (PBS-T), 0.01M. *p*H 7.2, containing 0.05%(v/v) Tween-20.

(iv)   **Substrate buffer** (25 mL) : Diethanolamine (10%, v/v) buffer, *p*H 9.6, containing substrate- *p*-nitrophenyl phosphate (1.25 mg/mL buffer) and 0.05 mM $MgCl_2$ (v) Sodium hydroxide 2 M.

Procedure

(i)   Coat the wells with 100 µL (100 pg - 500 ng of rabbit IgG) reference standard in coating buffer.

(ii)   Coat few wells with diluted normal rabbit serum (test samples, 100 µL/well) in coating buffer.

(iii) Coat few wells with 500 ng of BSA/100 µL in coating buffer, which serves as blank

(iv)   Dry the plate overnight at 37°C in an incubator.

(v)   Wash the plate with washing buffer, thrice and blot with tissue paper.

(vi)   Block the wells with 150 µL/well with blocking buffer and incubate for 30 min at 37°C.

(vii) Wash the plate with washing buffer, thrice and blot with tissue paper.

(viii) Add, 100 µL /well of anti-rabbit IgG raised in goat and labelled with alkaline phosphatase (1: 2500 diluted), in phosphate buffer (10 mM, *p*H 7.2) and incubate the plate for 1hour at 37°C.

(ix)   Wash the plate with washing buffer, thrice and blot with tissue paper.

(x)   Add, 150 µL/well of substrate buffer and incubate the plate for 20 -30 min at 37°C.

(xi)   Stop the enzyme reaction by adding 100 µL of 5 M NaOH per well.

(xii) Visually, compare the intensities of the yellow colour formed in the standards and the test samples (Fig. 12.6 & 12.7). Alternatively, record the absorbance at 405 nm in a spectrophotometer, using a micro-cuvette (0.5 mL volume) or in an ELISA reader and compute the concentration of the analyte in the test sample from the calibration plot (Fig. 12.8).

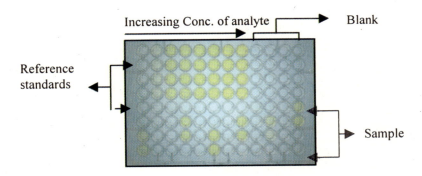

Fig. 12. 6. A typical antibody capture assay using (8 x 12 well) ELISA plate

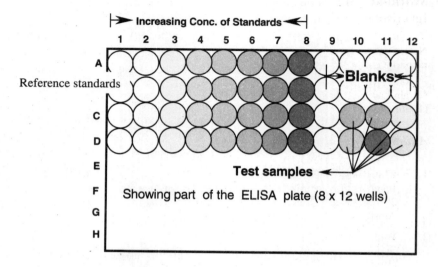

Fig. 12.7 Diagrammatic illustration of colour development
in an ELISA plate.

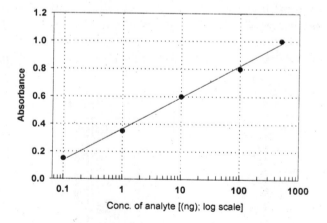

Fig. 12.8 A typical calibration curve of antibody capture assay

Workout: Determine the cross-reactivity of the anti-serum against bovine IgG (isolated in experiment #4, of this chapter).

Note: Coat separately, the ELISA plate with various concentration of bovine IgG instead of rabbit IgG.

$^\$$ Brain Law, (ed), Immunoassay: *A Practical Guide*. Taylor & Francis, Ltd. London, UK, (1996)

Additional Reading

1. Clausen, J. Laboratory Technique in Biochemistry and Molecular Biology: Immunochemical techniques for the identification and estimation of macromolecules. 2rd revised ed. Elsevier /North-Holland Biomedical Press, Science Publishers, Amsterdam, The Netherlands, (1981).

2. Weir, D. M., Herzenberg, L. A., & Blackwell, C. (eds), Hand Book of Experimental Immunology, 4th ed. Vol. 1, Blackwell Scientific Publications, Boston, MA,USA, (1986).

3. Harlow, E. & Lane, D. Antibodies: A Laboratory Manual. Cold Spring Harbor Laboratory, Cold Spring Harbor, New York, USA, (1988).

4. Kerr, M. A. & Thorpe, R., (eds), Immunochemistry LABFAX. BIOS, Scientific Publishers Ltd. Oxford, UK, (1994).

5. Lefkobits, I., (ed), Immunology Methods Manual: The Comprehensive Source Book of Techniques. Academic Press Inc. San Diego, CA, USA. (1996).

6. Pound, J. D., (ed), Immunochemical Protocols, 2nd ed. Vol. 80. Humana Press, New Jersey, USA. (1998).

Lab Notes

13

Food Analysis

Food is one of the basic necessities of human life. Consumption of nutritious food not only fulfills the energy needs of an individual, but is also a source of vital nutrients required for normal growth, development and healthy life. Human food requirements are mainly derived from plant and animal sources. Important foods derived from plant sources include cereals, millets, pulses and oilseeds, while foods like milk, eggs and flesh products are of animal origin. Nutritionally, the biochemical components of food are referred to as proximate principles. Major proximate principles of food include proteins, fats and carbohydrates, which form the bulk of human diet. Other important nutrients of food include vitamins and minerals that are required for normal metabolic activities of the body.

Some plant foods, besides nutrients, also contain anti-nutritional factors. These anti-nutritional factors interfere in the assimilation and utilization of nutrients contained in the foods. Common anti-nutritional components in the Indian food include, thermolabile enzyme inhibitors, such as trypsin, chymotrypsin and amylase inhibitors present in certain legumes and egg white phytates and phenolics/tannins present in cereals and millets and goitrogens present in oil seeds and legumes. Tannins combine with proteins to form tannin-protein complex, thus, causing an interference in protein digestibility.

More importantly, the food consumed by man has to be both nutritious and safe. Consumption of contaminated food leads to the spread of food borne diseases. Thus, food is the most complex part of the environment through which human population is exposed to a variety of contaminants and adulterants. These contaminants could either be naturally occurring or man-made chemicals. Food contamination can occur during production (pre-harvest, post-harvest), storage, transportation and processing of food. Contaminants of microbial origin include microbial toxins (ex. *Staphylococcus* toxins, botulin toxin), mycotoxins (ex. aflatoxins, trichothecenes) etc., while pesticides and heavy metals comprise common chemical contaminants of food. Conventional food adulterants include starch in milk, argemone oil in edible oils, lead chromate in turmeric, kesari dal (*Lathyrus sativus*) in pulses.

Control over food safety & hygiene is an integral part of any national welfare programme. The Prevention of Food Adulteration Act (1954) is the national legislation for ensuring food safety and protecting consumers against the supply of contaminated or adulterated foods.

Analytical procedures and techniques have been routinely used in the analysis of foods for the determination of (i) Proximate composition (ii) Anti-nutritional factors (iii) Vitamin and mineral composition (iv) Food contaminants & adulterants and (v) Food additives.

Procedures for few experiments on qualitative detection of food adulterants and quantitative analytical methods for the determination of proximate principles and anti-nutritional factors are detailed in this chapter.

(I) QUALITATIVE ANALYSIS OF SOME COMMON FOOD ADULTERANTS

Qualitative tests for the detection of adulterants in foods such as oils & fats, milk, milk products and artificial milk, pulses, spices & condiments, beverages are given below; these procedures include simple solubility and chemical tests.

Oils & Fats

Adulteration of edible oil by mineral oil, castor oil and argemone oil can be detected by performing the following tests:

Food component	Adulterant	Detection test
Edible Oil	Mineral oil	To 2 mL of edible oil, add equal volume of 0.5 N ethanolic potassium hydroxide and heat the contents in a boiling water bath for 15 minutes. Cool and add 10 mL of distilled water. Appearance of turbidity indicates the presence of mineral oil.

Edible oil	Castor oil	To 0.5 mL of the oil sample, add 2 mL of petroleum ether in a test tube and chill (0 to -5° C) the contents for 5 minutes. Appearance of turbidity within 5 minutes indicates the presence of castor oil.
Edible oil	Argemone oil	To 1 mL of oil sample, add few drops of concentrated nitric acid and vortex the contents. Development of red to reddish brown colour in the acid layer is indicative of argemone oil adulteration.

Milk, Milk products and synthetic milk

Food component	Adulterant	Detection test
Milk	Starch	To 1mL of milk sample add few drops of iodine reagent*. Blue colour confirms the presence of starch.
Milk Products	Hydrogenated fat	To 2 mL of clarified butter or ghee, add equal proportion of concentrated hydrochloric acid and a pinch of sucrose. Shake and leave the contents to stand for 5 minutes. Appearance of maroon colour in the lower acid layer is indicative of the presence of hydrogenated fat.
Artificial Milk	Urea	To 2 mL of milk sample, add few drops of bromocresol blue (0.1%w/v) indicator reagent. Appearance of dark blue colour is indicative of urea as an adulterant in artificial or synthetic milk.
Artificial Milk	Detergent	To 2 mL of milk sample, add few drops of bromocresol purple (0.1%w/v) indicator reagent. Development of a faint blue colour indicates the presence of detergent.

* See, *6. Qualitative analysis*, for iodine reagent preparation.

Beverages, spices, condiments:

Food component	Adulterant	Detection test
Tea leaves	Exhausted tea coloured with coal tar dye	Sprinkle few milligrams of tea dust onto a clean and white ceramic tile pre-coated with a layer of calcium hydroxide paste (slaked lime). Development of red-orange or other shades of colour indicates the presence of coal tar dye. In case of unadulterated tea, development of sparse greenish colour is due to the presence of chlorophyll.
Turmeric	(i) Sawdust coloured with coal tar dye. (ii) Lead chromate	To 1g of turmeric powder, add few drops of conc. HCl. Instant appearance of violet colour, which persists upon dilution with distilled water indicates the presence of sawdust coloured with a non-permitted coal tar dye (metanil yellow). Appearance of pink colour indicates the presence of lead chromate (a banned food colour).
Saffron	Coloured tendrils of maize cob (artificial saffron).	Sprinkle saffron sample on the surface of the distilled water (30 mL), contained in a 50 mL glass beaker. Instant appearance of colour indicates the presence of artificial dye. Pure saffron sample, when tested, will continue to give its colour as long as it lasts.
Asafoetida (Hing)	Soap stone or other earthy matter. Starch.	Take 0.5 g of the sample and add 3 mL of distilled water and vortex the contents. Soap stone or other earthy matter settles down at the bottom, indicating the presence of adulterants. In compounded hing, presence of starch is detected by the formation turbidity in the solution.

Pulses:

Food component	Adulterant	Detection test
Pulses	Kesari dal (*Lathyrus sativus*)	To 10 g of the sample, add 50 mL of 6 M HCl and boil contents with constant stirring, for 15 minutes on a hot plate. Appearance of deep pink colour confirms the presence of kesari dal.
	Lead chromate	To 5 g of pulse sample, add 10 mL of 2 M HCl. Development of pink colour shows the presence of lead chromate.

Workout: Collect random market samples (oil, milk & milk products, beverages, spices, condiments & pulses) and test for the presence of adulterants.

(II) QUANTITATIVE ANALYSIS

Proximate analysis: This term refers to the determination of the major constituents of food. These constituents include parameters such as (i) moisture (ii) fat (iii) ash and (iv) crude protein.

Experiment # 1: Determination of moisture content [§]

Principle: Food sample is dried at 125°C for 3h in a hot air oven. The loss in weight is reported as percent moisture.

Note: In case of cereals, heat for 1h at 130°C.

Equipment: Hot air oven with air circulation, analytical balance.

Glassware: Desiccator containing silica gel or fused calcium chloride, moisture dish with lid (made of glass or aluminum).

Procedure:

(i) Weigh accurately, 4-5 g of food commodity and transfer it into a pre-weighed moisture dish. The sample should be evenly spread.

(ii) Dry the sample for 3h at 125°C in a hot air oven. After drying, place the lid of the moisture dish and transfer the dish into a desiccator to cool the contents.

(iii) Weigh the moisture dish rapidly, after 30 min of desiccation and record the weight. Calculate the % moisture content by the following equation:

Calculation:

$$\% \text{ moisture} = \frac{(b-c) \times 100}{a}$$

Where,   *a*   = Weight of the sample in grams.
         b   = Weight of the dish + sample prior to drying.
         c   = Weight of the dish + sample after drying.
         (*b*– *c*) = Drop in weight due to loss of moisture.

Workout: Determine the moisture content of (i) groundnut seeds (ii) wheat flour.

§ FAO Food & Nutrition Papers: 14/7: Manuals of Food Quality Control, FAO, UN, Rome, Italy. 1986.

Experiment # 2: **Determination of oil content in oil seeds** (*Soxhlet* method)*§*

Principle: Oil content is determined in dried and powdered foods by solvent extraction using a *Soxhlet* apparatus. The solvent used to extract the oil is removed by evaporation and the residue is weighed and reported as percent oil.

Equipment: Analytical balance, mechanical blender / mixer-grinder, hot water bath, hot air oven, electric heating mantle, fume hood.

Materials: Whatman No. 42/ No. 2 or cellulosic thimble, porcelain chips.

Glassware: *Soxhlet* apparatus, evaporating dish, beaker.

Reagent: *n*-Hexane.

Procedure

(i)   Accurately weigh 6 g of the food sample and grind it into a fine powder, using a mechanical blender. Transfer the sample into a moisture dish and dry the sample at 125°C for 2 -3h.

(ii) Using a Whatman # 42 or 2 filter paper, make a pouch (3 x 10 cm) to hold the sample. Fold three sides of the pouch and securely staple the sides, so that no food particle escapes from the pouch. Accurately, weigh and transfer 5 g of the moisture free food sample into the pouch from the open end and staple it firmly. Alternatively, a commercially available cellulosic thimble can be used to hold the food sample. Cover the top of the thimble with clean absorbent cotton.

(iii) Assemble the *Soxhlet* extraction apparatus on a heating mantle. Place the pouch or thimble containing the food sample in the *Soxhlet* extraction tube. Pour about 160 -180 mL of hexane into the extraction flask along with few porcelain chips. Set the heating mantle temperature in such a way that at least 160 – 180 drops/min of solvents flow on to the pouch or the thimble. Continue the extraction for 6 - 8 h. After completion of the extraction, cool and disconnect the extraction flask. Transfer the contents of the flask quantitatively (with 2 or 3 repeated washings with hexane) into a clean, dry and pre-weighed 250 mL glass beaker (preferably, in two or three equal aliquots).

(iv) Place the beaker on a hot water bath and evaporate the solvent in a fume hood. Dry the residue in a hot air oven at 105°C for 30 minutes to remove traces of solvent and moisture. Cool the beaker in a desiccator, weigh the contents, and record the weight.

(v) Calculate the % oil content of the food sample by the following equation:

Calculation

$$\text{Oil content (\%)} = \frac{(a-b) \times 100}{c}$$

Where, a = Weight of the container + oil
b = Weight of the container
c = Weight of the sample in grams

Workout: Determine the oil content of (i) groundnut sample (ii) maize sample, procured from the market. Which sample contains more oil?

§ FAO Food & Nutrition Papers: 14/7: Manuals of Food Quality Control, FAO, UN, Rome, Italy. 1986.

Experiment # 3: Determination of ash content of foods [§]

Principle: The ash content of the food stuff represents its inorganic constituents after the organic and volatile material have been oxidised completely during the process of incineration at 600 °C in a muffle furnace.

Equipment: Muffle furnace, analytical balance.

Glassware: Desiccator, silica or porcelain crucible.

Procedure:

(i)   Weigh, 3 g of defatted food sample into a crucible of known weight.

(ii) Set the temperature of the muffle furnace to 600°C. Place the crucible containing the food sample into the muffle furnace after attaining the set temperature, using a metal tong. Incinerate the sample for 2h.

(iii) Transfer the crucible into a desiccator and cool to room temperature. Immediately, weigh the crucible along with ash, to prevent moisture absorption. Record its weight. Repeat the process of incineration until a constant weight is obtained. Calculate the percent ash content as follows:

$$\% \text{ Ash content} = \frac{(b-c) \times 100}{a}$$

Where, a = Weight of the sample in grams.
b = Weight of the crucible + constant weight of ash.
c = Weight of the empty crucible.
$(b-c)$ = Weight of ash.

Workout: Determine the ash content of defatted groundnut and maize sample.

[§] FAO Food & Nutrition Papers: 14/7: Manuals of Food Quality Control, FAO, UN, Rome, Italy. 1986.

Experiment # 4: **Colorimetric determination of crude protein (*Kjeldahl* nitrogen – N)** [§]

Principle: Crude protein present in the food sample is digested with sulphuric acid in the presence of a catalyst, at 380°C. The nitrogen released from the proteins and non-protein constituents of the sample is converted to ammonium sulphate. The ammonia nitrogen reacts with salicylate-nitroprusside reagent in the presence of sodium hypochlorite, to form a green coloured complex, whose absorbance is measured at 685 nm. The crude protein concentration is calculated by multiplying the concentration of nitrogen (*Kjeldahl* nitrogen - N) obtained with a factor of 6.25. (Crude protein = N x 6.25).

Equipment: Analytical balance, digestion unit, mechanical blender / mixer-grinder.

Reagents: (i) Salicylate reagent: Dissolve 3.2 g of sodium salicylate, 4.0 g of tri-sodium phosphate and 50 mg of sodium nitroprusside in warm water and make up the volume to 100 mL with distilled water. Store the reagent in an amber coloured bottle (ii) Hypochlorite reagent: Take 5mL of 5%(v/v) sodium hypochlorite solution and make up the volume to 100 mL with distilled water (iii) Sodium hydroxide solution (5 N) (iv) Conc. sulphuric acid.

Digestion mixture: Mix and grind thoroughly, 35 g of potassium sulphate and 0.35 g of selenium using a glass mortar and pestle.

Standard Ammonium chloride solution: Dissolve 38.2 mg of pure ammonium chloride in distilled water and make up the volume to 100 mL in a volumetric flask. This stock solution contains 100 μg equivalent of N per mL. Dilute the stock solution to obtain 20 μg/mL of nitrogen.

Procedure

(i)   Take 100 mg of the defatted food sample (defat the sample by Soxhlet method) in a *Kjeldhal* digestion flask. Add 2 g of the digestion mixture (catalyst) along with 2 mL conc. sulphuric acid and mix. Digest the contents at 380 °C for 1h, in a chemical safety hood. Cool and transfer the contents into a 100 mL volumetric flask and make up the volume with distilled water. Take suitable aliquots (ex. 0.05, 0.1, 0.2 mL or appropriate dilution) to determine the nitrogen content in the sample.

Note: If no colour develops, neutralize the acidity by adding 0.1 mL of 5N NaOH.

(ii)   Run a set of standards in the range of 4 - 20 µg N, as given in the protocol below. Construct a calibration curve by plotting concentration of nitrogen on x - axis and absorbance at 685 nm on y -axis. Compute the concentration of nitrogen in the sample from the standard curve. Calculate the crude protein content in the sample by the following equation:

$$\text{Crude protein content } (\%) = N \times 6.25*$$

* **Note**: Following factors are used for converting N to crude protein in various foods, like (i) Milk and milk products - N x 6.33 (ii) ground nuts – N x 5.46 and (iii) Cereals – N x 6.25.

Protocol:

S.No	Std. solution (mL)	Conc. of Nitrogen (µg)	Distilled water (mL)	Salicylate reagent (mL)	Hypochlorite reagent (mL)	Absorbance (685 nm)
1.	Blank	- -	11.0	4.0	1.0	
2.	0.2	4.0	10.8	"	"	
3.	0.4	8.0	10.6	"	"	
4.	0.6	12	10.4	"	"	
5.	0.8	16	10.2	"	"	
6.	1.0	20	10.0	"	"	
7.	Sample	(to be determined)	(Appropriate vol.)	"	"	

Workout: Determine the nitrogen content of (i) Defatted groundnut (ii) Casein. Calculate the crude protein content.

§ Willis, R. B. Montgomery, M. E. & Allen, P. R. J. Agric. Food Chem. 44: 1804, (1996).

Experiment # 5: Quantitation of non-protein nitrogen (NPN)[§]

The non-protein nitrogen in foods includes nitrogenous bases of nucleic acids, non-protein amino acids (ex. Ornithine, homocysteine, citrulline, β-alanine, diamino propionic acid, desmosine, etc., amino sugars and their derivatives, phospho- and sphingolipids etc.).

Principle: Proteins present in food sample are precipitated by trichloroacetic acid (TCA) and the non-precipitated nitrogenous components in the supernatant are estimated by salicylate-nitroprusside method.

Reagent: Trichloroacetic acid (10% w/v).

Procedure

(i)   Weigh 1 g of defatted food sample and transfer it into a 20 mL graduated and stoppered centrifuge tube. Add 15 mL of TCA reagent and mix for 1h in a wrist action shaker. After 1h remove the stopper and centrifuge the contents at 1500 rpm for 15 minutes. Carefully collect the supernatant into a 25 mL volumetric flask. Wash the residue twice with 5 mL TCA reagent and repeat the centrifugation step. Pool the TCA extracts and make up the final volume to 25 mL with TCA reagent.

(ii)   For the determination of non-protein nitrogen, take aliquots of 5 and 10 mL of TCA extracts. Determine the nitrogen content of the sample after digesting the extract with sulphuric acid as described in experiment #4.

(iii) Express the value of non-protein nitrogen as percent.

Workout: Determine the non-protein nitrogen content of defatted groundnut sample. Express the NPN value as percent of the total nitrogen.

[§] Singh, U., & Jambunathan, R. J. Agric. Food Chem. 29: 423, (1981).

Analysis of anti-nutritional factors, rancidity and iron in foods

Experiment # 6: Determination of trypsin inhibitor activity (TIA)[§]

Principle: Activity of trypsin in the presence and absence of the inhibitor is assayed using casein as an enzyme substrate. The enzymatic breakdown of casein is spectrophotometrically monitored at 280 nm.

Equipment: Analytical balance, *p*H meter, spectrophotometer, constant temperature water bath, steam bath, wrist action shaker.

Reagents: (i) Sodium phosphate buffer (100 mM, *p*H 7.6) (ii) Substrate solution (2% w/v casein): Dissolve 2 g of casein in 80 mL phosphate buffer by heating over a steam bath for 15 min. Cool and make up the volume with phosphate buffer to 100 mL in a volumetric flask (iii) Hydrochloric acid (0.001M) (iv) Trichloroacetic acid (5% w/v).

Stock trypsin solution: Dissolve 5 mg trypsin (~200 Units/mg solid) in 100 mL of 0.001 M HCl. (stable for 2 weeks, if stored in refrigerator).

Sample: Defatted soybean flour.

Procedure

(i)   Weigh 250 mg of defatted soybean flour in a stoppered 100 mL Erlenmeyer flask. Add 10 mL of 100 mM phosphate buffer and the contents are shaken using a wrist action shaker at low speed (avoid frothing of the sample) for 1h at room temperature. Transfer the sample into a 15 mL capacity conical centrifuge tube and centrifuge at 5000 x g for 5 min. Collect the supernatant, which is a source of soybean trypsin inhibitor protein (also known as *Kunitz* protein).

(ii)   Trypsin enzyme assay: Dispense 0.2, 0.4, 0.6, 0.8 and 1.0 mL of stock trypsin solution into a duplicate set of labeled test tubes. Make up the volume in each test tube to 2.0 mL by adding phosphate buffer. Place the tubes in a constant temperature water bath at 37°C for 5 minutes. Add 6 mL of 5% TCA to one set of test tubes containing different volumes of enzyme, which serves as blank. Add 2 mL of substrate solution to all the tubes and note the time. Incubate the tubes for exactly 20 min. After 20 min, stop the enzymatic reaction in the experimental tubes by adding 6 mL of 5% TCA. Allow the test tubes to stand at room temperature for 1h and centrifuge at 3000 x g for 10 min. Read the absorbance of the supernatant at 280 nm against respective blanks.

(iii) Assay of trypsin inhibitor activity: Dispense, 0.2, 0.4, 0.6, 0.8 and 1.0 mL of sample extract into a duplicate set of labeled test tubes. Make up the volume in each test tube to 1.0 mL by adding phosphate buffer. Add 1 mL of stock trypsin solution to all test tubes and place them in a constant temperature water bath, at 37°C, for 5 min. Add 6 mL of 5% TCA to one set of test tubes containing different volumes of enzyme and sample extract, which serves as blank. Add 2 mL of substrate solution to all the tubes and note the time. Incubate the tubes for exactly 20 min. After 20 min, stop the enzymatic reaction in the experimental tubes by adding 6 mL 5% TCA. Allow the test tubes to stand at room temperature for 1h and centrifuge at 3000 x g for 10 min. Read the absorbance of the supernatant at 280 nm against respective blanks.

Calculation: The difference in absorbance at 280 nm due to breakdown products formed by a given concentration of trypsin in the presence and absence of soybean trypsin inhibitor gives a measure of trypsin inhibitor activity. One trypsin unit (TU) is arbitrarily defined as an increase of 0.01 absorbance value at 280 nm, in 20 min at 37°C per 10 mL volume of reaction mixture. Trypsin inhibitor activity is expressed as number of trypsin units inhibited per mg of defatted sample or per mL of the sample extract. Plot a linear graph between $(A^{2\,3})$ - a mathematically transformed A_{280nm} value $(A^{2\,3} = \sqrt[3]{A^2})$ on y – axis vs range of trypsin enzyme concentration used in the analysis, on x-axis. This mathematical transformation of absorbance value is employed to convert curvilinear response to linear one.

Workout: Find out the consequence of heat treatment on soybean extract.

Procedure: Treat 10 mL of the soybean extract at 85°C for 30 min and assay the trypsin inhibitor activity.

§Kakade, M. L., Simons, N., & Liener, I. E. Cereal Chem. 46: 518, (1969).

Experiment # 7: **Estimation of polyphenols/tannins by *Folin-Denis* method** [§]

Principle: *Folin-Denis* reagent oxidizes polyphenolic compounds present in food commodities to blue colour, which is measured at 760 nm in a spectrophotometer. The intensity of the colour is proportional to the tannin content.

Equipment: Analytical balance, electric heating mantle and hot water bath.

Glassware: All glass refluxing unit.

Reagents: (i) *Folin-Denis* reagent: Dissolve in a refluxing flask (2 L capacity), 100 g sodium tungstate and 20 g of sodium molybdate in 750 mL of distilled water. Add 50 mL of ortho-phosphoric acid (85-88% v/v) and reflux the contents for 8 to10 h. Cool the flask to room temperature and make up the volume to 1 L. Store the reagent in an amber coloured reagent bottle (ii) saturated sodium carbonate solution: To 100 mL of distilled water taken in a glass beaker, add 45 grams of anhydrous sodium carbonate. Heat the contents on a hot water bath at 70 - 80°C, with intermittent stirring to dissolve. Cool the contents overnight and seed this solution by adding few crystals of sodium carbonate. Filter this reagent using a funnel plugged with glass wool (iii) Extraction solvent: Methanolic-HCl (1% (v/v) of conc. HCl in methanol) .

Tannic acid standard: Weigh accurately 10 mg of tannic acid into a 100 mL volumetric flask. Dissolve and make up the volume with distilled water (to be freshly prepared).

Procedure

(i)   Weigh 250 mg of defatted food material into a 250 mL round bottom flask. Add 100 mL of extraction solvent and reflux for 2 h over a heating mantle. Allow the contents to cool and filter the extract quantitatively into a 100 mL volumetric flask. Make up the volume with methanolic-HCl (sample extract can be stored, refrigerated for 2 days).

(ii)   Dispense 0.1 to 1 mL aliquots of tannic acid standard (range 10 -100 µg) into labeled test tubes. Add to each test tube 7.5 mL distilled water. Simultaneously run a reagent blank. Take separately 0.05, 0.1 and 0.2 mL of sample extract into labeled test tubes and add 7.5 mL of distilled water.

(iii) Add 0.5 mL of *Folin-Denis* reagent to all test tubes and vortex the contents. This is followed by the addition of 1 mL of saturated sodium carbonate solution. Make up the final volume to 10 mL with distilled water. Mix and allow tubes to stand at room temperature for 30 min and record the absorbance at 760 nm immediately. Construct a calibration curve and compute the concentration of polyphenols in the sample. Express the value as mg tannic acid equivalent per gram of sample.

Workout: Determine the tannin content in defatted (i) Sorghum (*Sorghum bicolor*) sample and (ii) Groundnut sample.

[§] (i)   Folin, O., & Denis, W. J. Biol. Chem. 12: 239, (1912).
(ii) AOAC, Official Methods of Analysis. 14th ed, AOAC, Arlington, VA, USA. (1984).

Experiment # 8: Determination of rancidity in edible oil: *Kries* **test** [§]

Principle: Development of rancidity in oils is a complex process brought about by the action of (i) air (-oxidative rancidity) or (ii) microorganisms (-ketonic rancidity). Rancid oil reacts with phloroglucinol (1,3,5-trihydroxybenzene) under acidic condition to yield a pink colour, which is measured at 545 nm.

Phloroglucinol

Reagents: (i) Acid reagent (100 mL): Trichloroacetic acid (30 w/v) in glacial acetic acid (ii) Phloroglucinol (1%w/v) in glacial acetic acid (iii) Ethanol (95% v/v) and (iv) Chloroform.

Procedure: Take 5 mL of test oil sample and add 5 mL of chloroform in a stoppered boiling test tube and mix. Add 10 mL of acid reagent followed by 1 mL of phloroglucinol reagent. Incubate the tube at 45°C in a hot water bath for 15-18 minutes. After incubation add 4 mL of ethanol (95% v/v) and immediately record the absorbance at 545 nm. Run the test in triplicates.

Interpretation of test results:

(i)   An absorbance value of ≥1.0 indicates that the sample is highly rancid.
(ii) An absorbance value of > 0.2 indicates that the sample has incipient rancidity.
(iii) An absorbance value ≤ 0.15 indicates no rancidity.

Note: Development of rancidity in oils is a complex process, which needs more than one test for confirmation. Normally, rancidity test include determination of (i) free fatty acids (ii) peroxide value (iii) *Kries* value.

Workout: Test the rancidity in (i) fresh coconut oil (*Cocos nucifera*) (ii) stored /aged safflower oil (*Carthamus tinctorius*) and (iii) repeatedly used groundnut oil.

§ (i) Bedford, C. L., & Joslyn, M. A. Food Res. 2: 455, (1937).
  (ii) Tappel, A. L., Knapp, F. W., & Urs, K. Food Res. 22: 287, (1957).

Experiment # 9: Determination of iron content in apple juice §

Principle: Iron present in biological sample is released by acid treatment, which undergoes reduction to ferrous state in the presence of thioglycolic acid. This iron complexes with bathophenanthriline (1,10-phenanthroline) reagent to yield a pink colour, whose intensity is measured at 535 nm.

Reagents: (i) Protein precipitant reagent: Dissolve 100 g of trichloroacetic acid (TCA) in 250 - 300 mL in deionized or double glass distilled water. Separately mix 30 mL of thioglycolic acid and 2 mL of conc. HCl, add it to TCA solution and make up the volume to 1 L in a volumetric flask with deionized or double glass distilled water (ii) Chromogen reagent: Weigh 25 mg phenanthroline, dissolve and make up the volume to 100 mL with 2 M sodium acetate solution (prepared in deionized or double glass distilled water) in a volumetric flask.

Note: Ensure all the glassware used in the experiment is free of any iron contamination.

Sample: Apple juice (use any commercially available product, which may be procured from market).

Standard iron solution: Stock solution (100 μg of iron/mL): Dissolve 70.22 mg of ammonium ferrous sulphate [$(NH_4)_2SO_4FeSO_4 \cdot 6H_2O$] in double glass distilled water and make up the volume to 100 mL. Dilute the stock solution 1:10 in double glass distilled water to give a final concentration of 10 μg of iron/mL.

Procedure

(i) Take 1 mL of apple juice in an iron-free test tube, add 1 mL of protein precipitant reagent and vortex for 1 minute. Allow the sample to stand at room temperature for 5 minutes and centrifuge at 2500 x g for 15 minutes.

(i) Run a set of iron standards separately in the range of 1- 10 μg mL^{-1} and a reagent blank, along with sample. Add 1 mL of protein precipitant reagent to all tubes.

Protocol

S.No	Vol. of standard solution (mL)	Concentration of iron (μg)	Vol. of distilled water (mL)	Precipitant reagent (mL)	Chromogen reagent (mL)	Absorbance at 535 nm
1.	Blank	- -	2.0	1.0	1.0	
2.	0.1	1.0	1.9	,,	,,	
3.	0.2	2.0	1.8	,,	,,	
4.	0.4	4.0	1.6	,,	,,	
5.	0.8	8.0	1.2	,,	,,	
6.	1.0	10.0	1.0	,,	,,	
7.	Sample	To be determined	(Appropriate vol.)	,,	,,	

,, = same volume

(ii)   Carefully collect and transfer the supernatant of the test sample into a separate test tube. Add 1 mL of chromogen solution to test tubes containing blank, standards and test solutions. Mix and allow the test tubes to stand for 10 – 12 minutes at room temperature before recording the absorbance at 535 nm in a photometer. Construct a calibration curve and compute the concentration of iron in the apple juice. Express the value per 100 mL of the juice.

Workout: Determine the iron content in (i) Watermelon juice (ii) Grape juice. Find out the iron-rich fruit juice among the samples tested (apple, watermelon and grape juice) and rank them.

§ Smith, F. G. *et al., Analyst.* 77 : 418, (1952).

Additional Reading

1.   Gruenwedel, D. W. & Whitaker, J. R. *Food Analysis*: Principles and Techniques. Vol. 1 & 2, Marcel Dekker: New York (1984).

2.   Kirk, R. S. & Sawyer, R. *Pearson's* Chemical Analysis of Foods. 9th ed. Longman Scientific and Technical; Harlow, Essex, UK (1991).

3.   AOAC International. 'Official Methods of Analysis'. 16th ed, AOAC International, Gaithersburg, MD, USA (1995).

4.   Coultate, T. P. *Food* - The Chemistry of its Components. 3rd ed, Royal Society of Chemistry, London, UK (1996).

5.   Miller, D. D. *Food Chemistry*: A Laboratory Manual. John Wiley & Sons Inc. New York, USA (1998).

Appendix

A. Physical constants

Physical constants	Symbol	Value
Acceleration due to gravity	g	9.81 m s^{-2}
Charge/mass ratio	e/m	1.758796×10^{11} C* kg^{-1}
Electronic charge	e	1.60219×10^{-19} C
Planck constant	h	6.6262×10^{-34} J s^{-1}
Atomic mass unit	m_u	1.660566×10^{-27} kg
Speed of light in vacuum	c	2.997925×10^{8} m s^{-1}
Avogadro constant	N_A	6.02205×10^{23} mol^{-1}
Gas constant	R	8.314 J K^{-1} mol^{-1}

* C = charge containing 6.28×10^{18} electrons

B. Visible light absorption, filters & complementary hues

Wavelength (nm)	Filter (Colour absorbed)	Complementary hue (Colour transmitted)
380-435	Violet	Yellowish green
436-480	Blue	Yellow
481-490	Greenish blue	Orange
491-500	Bluish green	Red
501-560	Green	Purple
561-580	Yellowish green	Violet
581-595	Yellow	Blue
596-650	Orange	Greenish blue
651-780	Red	Bluish green

C. Boiling point & density of some organic solvents routinely used in the laboratory

Solvent	Boiling point (°C)	Density*
Acetone	56	0.791
Acetonitrile	82	0.782
Acetic acid	118	1.490
Acetic anhydride	136	1.082
Aniline	184	1.022
Benzene	80	0.879
n - Butanol	118	0.810
sec - Butanol	100	0.808
Chloroform	61	1.480
Carbontetrachloride	77	1.594
Carbondisulphide	46	1.263
Diethyl ether	35	0.714
Ethyl alcohol	78	0.791
Ethylene glycol	197	1.109
Ethylene glycol mono-methyl ether (Methyl Cellosolve)	135	0.930
Ethyl acetate	77	0.901
Glycerol	290	1.260
n-Hexane	69	0.649
iso-Propanol	82	0.785
Methyl alcohol	65	0.792
Pyridine	115	0.982
Phenol**	182	1.071
Toluene	111	0.867

* - at 20° C; **Melting point = 43° C

D. Molar absorption coefficients of selected biochemicals

Biomolecule	Absorption maxima (λ_{max}) (nm)	Molar absorption coefficient (ε x 10^{-3})
Phenylalanine	257.5	0.19 (in 0.1N HCl)
	258.0	0.21 (in 0.1N NaOH)
Tryptophan	278.0	5.60 (in 0.1N HCl)
	280.5	5.43 (in 0.1N NaOH)
Tyrosine	274.5	1.34 (in 0.1N HCl)
	293.5	2.33 (in 0.1N NaOH)
NAD$^+$ & NADP$^+$	259	18.0*
NADH & NADPH	339	6.22*
2,4-Dinitrophenol	400	18.3 (at alkaline pH, 9-10)
Riboflavin	450	12.2*
Adenine	260	13.3*
Guanine	260	7.2*
Cytosine	260	2.6*
Thymine	260	7.4*
Uracil	260	8.2*
ATP	260	15.4*
GTP	260	11.7*
DNP-alanine	360	17.2 (in 1 N NaOH)
DNP-aspartic acid	360	18.2 (in 1 N NaOH)
DNP-tyrosine	360	16.9 (in 1 N NaOH)

* at pH 7

E. Molarity, normality,% concentration and specific gravity of some acids & alkalies

	MW	Molarity	Normality	% Concentration	Specific gravity
Acid*					
Hydrochloric acid	36.46	11.6	11.6	36	1.18
Sulphuric acid	98.08	17.8	36.0	97	1.84
Nitric acid	63.02	16.4	15.7	71	1.42
o- phosphoric acid	98.00	14.7	44.0	85	1.71
Acetic acid	60.00	17.5	17.5	99	1.06
Alkali					
Potassium hydroxide	56.00	11.7	11.7	45	-
Sodium hydroxide	40.00	10.0	10.0	40	-
Ammonium hydroxide (Conc.)	17.00	-	15.0	28**	0.90

* Concentrated acids; ** with respect to ammonia; **MW** = Molecular weight

F. Preparation of acid & alkali (1 N) solutions

	Acid/Alkali	Final volume to be made up to one liter with distilled water
1.	Concentrated sulphuric acid	30 cc
2.	Concentrated hydrochloric acid	100 cc
3.	Concentrated nitric acid	63 cc
4.	Oxalic acid dihydrate	62.96 g
5.	Sodium hydroxide	40.00 g
6.	Potassium hydroxide	56.00 g

G. *p*H indicators for volumetric analysis

Indicator	Colour in		*p*H range
	Acid	Base	
Methyl orange	Red	Yellow	3.1 - 4.4
Methyl red	Red	Yellow	4.4 - 6.2
p-nitrophenol	Colourless	Yellow	5.6 - 7.6
Phenol red	Yellow	Red	6.4 - 8.2
Phenolphthalein	Colourless	Pink	8.2 - 10.0

H. Standard buffers for *p*H meter calibration

Chemical	Molarity	*p*H	Temperature (^oC)
Potassium hydrogen phthalate	0.05	4.00	25
		4.01	30
		4.02	35
Phosphate buffer	0.05	6.86	25
		6.85	30
		6.84	35
Sodium tetraborate (Borax)	0.01	9.18	25
		9.14	30
		9.10	35

PREPARATION OF STANDARD BUFFER SOLUTIONS

1. Potassium hydrogen phthalate (0.05M): Dissolve 1.021 g of phthalate in distilled water and make up the volume to 100 mL in a volumetric flask

2.   Phosphate buffer (0.05M): Dissolve 340 mg of potassium phosphate -monobasic (KH_2PO_4) and 355 mg of anhydrous sodium phosphate - dibasic (Na_2HPO_4), in distilled water and make up the volume to 100 mL, using a volumetric flask.

3.   Borax (0.01M): Dissolve 383 mg of borax in distilled water and make up the volume to 100 mL using a volumetric flask.

I. *p*H values (approximate value) of frequently used 0.1M solutions

Solution	*p*H*
Ammonium hydroxide	11.3
Ammonium oxalate	6.4
Ammonium sulphate	5.5
Borax	9.2
Oxalic acid	1.3
HCl	1.1
Trichloroacetic acid	1.2
Potassium acetate	9.7
Sodium acetate	8.9
Sodium bicarbonate	8.3
Sodium carbonate	11.5
Sodium hydroxide	12.9
Sodium phosphate (monobasic)	4.5
Sodium phosphate (dibasic)	9.2

* at 25°C

J. Amino Acids: Abbreviation, molecular weight, *p*I value, & solubility.

Amino acid	Single letter & Triple letter code		Molecular weight	*p*I* value	Solubility (mg/ml of H_2O)
Alanine	A	Ala	89	6.107	158.0
Arginine	R	Arg	179	10.76	718.0
Asparagine	N	Asn	132	5.410	24.0
Aspartic acid	D	Asp	133	2.980	4.2
Cysteine	C	Cys	121	5.020	Freely
Glutamic acid	E	Glu	147	3.080	7.2
Glutamine	Q	Gln	146	5.650	26.0
Glycine	G	Gly	75	6.064	225.0
Histidine	H	His	155	7.640	41.9
Isoleucine	I	Ile	131	6.038	33.6
Leucine	L	Leu	131	6.036	23.7
Lysine	K	Lys	146	9.470	666.0
Methionine	M	Met	149	5.740	51.4
Phenylalanine	F	Phe	165	5.910	27.0
Proline	P	Pro	115	6.30	154.0
Serine	S	Ser	105	5.680	362.0
Threonine	T	Thr	119	5.870	Freely
Tryptophan	W	Trp	204	5.880	10.6
Tyrosine	Y	Tyr	181	5.630	0.38
Valine	V	Val	117	6.002	56.0

* at 25°C

K. Molecular weight of certain proteins

Protein	Molecular weight (kDa)
Lysozyme	14.3
Myoglobin	17
Trypsinogen	24
Carbonic anhydrase	29
Egg albumin	45
Bovine serum albumin (BSA)	66
Haemoglobin (Hb)	68
Immunoglobulin G (IgG)	150
Catalase	240
Myosin	205
Thyroglobulin	670

L. Isoelectric point of few proteins

Protein	Isoelectric pH
Pepsin	1.0
Egg albumin	4.6
Casein	4.8
BSA	4.9
Gelatin	4.8
Urease	5.0
Haemoglobin	6.8
Myoglobin	7.0
Ribonuclease	9.6
Lysozyme	11.0

M. Determination of protein concentration by UV- absorption method

Pure protein solutions can be quantitated by recording the absorbance at 280nm (A_{280}) or 205 nm (A_{205}). The absorbance of proteins at 280 nm is mainly due to tryptophan and tyrosine residues, while absorbance of proteins in the region of 205-220 nm is primarily due to peptide bonds, even though some amino acids may also contribute to this absorbance. This is a rapid method of protein quantitation in solution, with no destruction of the sample during analysis. The concentration of the protein (mg mL^{-1})^s is calculated by the following equations:

$$A_{205} \div 27.0 + \left[\frac{A_{280}}{A_{205}} \right]^* \qquad \dots\dots\dots Equation\ 1$$

$$[(1.55 \times A_{280\,nm}) - (0.76 \times A_{260\,nm})] \qquad \dots\dots\dots Equation\ 2$$

Note: * Nucleic acid contamination in the sample may lead to inaccurate values for proteins.

^{s} (i) Bailey, J. L., *Techniques in Protein Chemistry*, Elsevier, NY, USA, (1967).
  (ii) Paterson, G. L., *Methods Enzymol.* 91: 95-119, Academic Press, NY, USA, (1983).

UV-absorbance value of few reference proteins

Protein concentration (mg mL^{-1})*	Absorbance at 280 nm
Bovine serum albumin (BSA)	0.70
Immunoglobulin G (IgG)	1.35
Egg yolk antibody (hen-IgY)	1.23
Egg albumin	0.82
Casein	1.20

*At alkaline *p*H

N. Saponification, iodine & acid values of common oils & fats

Oil/Fat	Saponification No.	Iodine No.	Acid value
Beef tallow	196 - 200	35-42	0.25
Butter	210 - 230	26-28	0.46 - 35.0
Castor oil	175 - 183	82-90	0.10 - 0.80
Coconut oil	253 - 262	6-10	1.10 - 1.90
Corn oil	187 - 192	120-144	0.1 - 0.13
Cotton seed oil	194 - 196	103-111	0.60 - 0.90
Groundnut oil	188 - 196	84-104	0.15 - 0.50
Linseed oil	188 - 195	175-200	0.10 - 0.30
Mustard oil	185 - 195	105-126	0.10 - 0.30
Olive oil	185 - 196	79-88	0.30 - 1.00
Palmoline oil	195 - 205	54-62	0.06 - 0.10
Safflower oil	185 - 195	140-150	0.05 - 0.30
Sesame (Til) oil	188 - 183	103-120	≤ 0.06
Soya bean oil	192 - 195	127-143	0.05 - 0.15
Rice bran oil	188 - 193	103-120	0.05 - 0.10

O. DETERMINATION OF NUCLEIC ACID CONCENTRATION BY UV ABSORPTION METHOD

Pure nucleic acid solution shows absorption maxima at 260 nm. This property is typically made use in the spectrophotometric determination of nucleic acids. For a nucleic acid solution with an $A_{260} = 1$, the following approximations are valid,

   1 A_{260} unit* of double stranded DNA   = 50 µg /mL
   1 A_{260} unit* of single stranded DNA   = 37 µg /mL
   1 A_{260} unit* of single stranded RNA   = 40 µg /mL
   1 A_{260} unit* of oligonucleotide   = 20 to 33 µg /mL

(which is dependent on the length and the base sequence of the oligonucleotide).

* **Unit definition**: The concentration of nucleic acid dissolved in 1 mL buffer (20 mM sodium phosphate, *p*H 7.0 or 0.1M NaCl), which has an absorbance of 1 (at A_{260}). The spectrophotometric measurement is made in a 1cm cuvette, at 20 °C.

Further, the ratio between A_{260} (λ_{max} for nucleic acid) and A_{280} (λ_{max} for protein) provides a good index of the purity of a nucleic acid preparation. In solution, pure DNA and RNA typically have A_{260}/A_{280}, ratio of 1.8 and 2.0, respectively. If the ratio is found to be less than 1.8 or 2.0, then it can be inferred that the nucleic acid preparation is contaminated with protein or phenol

P. Some normal human clinical chemistry values of blood, serum & urine

Biological Fluid			Normal Clinical Chemistry Values
Blood			
1.	*p*H		7.35 - 7.45
2.	Specific gravity		$1.056^{\S}$
3.	Hemoglobin		
	(Male)		14 - 18 g/dL
	(Female)		12 - 16 g/dL
4.	Urea		14 - 50 mg/dL
5.	Urea Nitrogen		8 - 20 mg/dL
6.	Glucose (Fasting)		60 - 90 mg/dL
	(Post-prandial)		90 - 120 mg/dL
Serum			
1.	**Specific gravity**		$1.0254 - 1.0288^{\S}$
2.	**Proteins**		
	(i)	Total proteins	5.5 - 8.9 g/dL
	(ii)	Albumin	3.5 - 5.5 g/dL
	(iii)	Globulin	2.0 - 3.5 g/dL

(Contd.)

	(iv)	IgG	800 - 1500 mg/dL
	(v)	A:G ratio	1.5 – 2.5
3.	**Lipids**		
	(i)	Lipids (total)	400 - 600 mg/dL
	(ii)	Cholesterol (total)	160 - 240 mg/dL
	(iii)	Triglycerides	50 - 150 mg/dL
	(iv)	Phospholipids	150 - 250 mg/dL
4.	**Metabolites (organic)**		
	(i)	Bilirubin total	0.3 - 1.0 mg/dL
		(Direct)	0.1 - 0.3 mg/dL
		(Indirect)	0.2 - 0.7 mg/dL
	(ii)	Creatinine	0.6 - 1.5 mg/dL
	(iii)	Uric acid	
		(Male)	2.0 - 7.0 mg/dL
		(Female)	2.0 - 6.0 mg/dL
5.	**Metabolites (inorganic)**		
	(i)	Calcium (ionized)	4.5 - 5.6 mg/dL*
		(total)	9.0 - 11.0 mg/dL*
	(ii)	Chlorides	95 - 106 mEq/L
	(iii)	Phosphorus (inorganic)	3.0 - 4.5 mg/dL
	(iv)	Potassium	3.5 - 5.0 mEq/L
	(v)	Sodium	135 - 145 mEq/L
	(vi)	Iron	50 - 175 μg/dL
	(vii)	Bicarbonate	24 - 30 mEq/L
6.	**Enzymes**		
	(i)	Amylase	60 - 180 units (Somogyi)/dL
	(ii)	Lactate dehydrogenase	25 - 100 IU/L
	(iii)	Acid phosphatase	1 - 5 units (King-Armstrong)/dL
	(iv)	Alkaline phosphatase (Adults)	5 - 13 units (King-Armstrong)/dL
	(v)	Alanine amino transaminase (ALT; SGPT)	3 - 26 IU/L
	(vi)	Aspartate amino transaminase (AST; SGOT)	6 - 18 IU/L

(Contd.)

(vii)	Creatine phosphokinase	
	(Male)	5 - 35 IU/L
	(Female)	5 - 25 IU/L
(viii)	Lipase	0.2 - 1.5 IU/L

Urine

(i)	Volume (24h)	600 - 2500 mL
(ii)	Specific gravity	1.003 - 1.030
(iii)	pH	4.8 - 8.5 (mean value 6.0)
(iv)	Creatine	100 mg/24h
(v)	Creatinine	1 - 2 g/24h
(vi)	Urea	6 - 30 g/24h
(vii)	Uric acid	0.2 - 0.7 g/24h
(viii)	Protein (Albumin)	0.025 - 0.070 g/24h
(ix)	Urea nitrogen	10 -15 g/24h
(x)	Urobilinogen	0 - 4 mg/24h
(xi)	17-Ketosteroids	
	(Male)	6 - 20 mg/24h
	(Female)	1 -16 mg/24h
(xii)	Calcium	100 - 300 mg/24h
(xiii)	Chloride	170 - 254 mEq/24h
(xiv)	Phosphorus (inorganic)	0.35 - 1.5 g/24h
(xv)	Titratable acidity (24h)	25 - 70 mEq

§ = Varies with protein concentration;

* For conversion to mEq/L =

$$\frac{\text{mg/dL (or mg\%) x 10 x valence}}{\text{Atomic weight}}$$

Q. Some useful Websites for Biochemistry/Cell biology/Immunology/ Chemical safety data

http://www.biochemist.com
http://www.worthington-biochem.com
http://www.ultranet.com
http://www.biochemistry.org
http://www.biochem.ucl.ac.uk
http://www.mcb.harvard.edu
http://www.gwu.edu
http://www.biology.arizona.edu
http://www.ncbi.nlm.nih.gov
http://www.antibodyresource.com
http://www.iubio.indiana.edu
http://www.msdssearch.com
http://www.instruct.cit.cornell.edu
http://www.brenda.uni.koeln.de
http://www.omega.cc.umb.edu
http://www.chemfinder.cambridgesoft.com
http://www.biosupplynet.com

Index